Integrated Food and Bioprocessing

Abdul Sattar Jatoi
Editor

Integrated Food and Bioprocessing

Springer

Editor
Abdul Sattar Jatoi
Department of Chemical Engineering
Dawood University of Engineering and Technology
Karachi, Pakistan

ISBN 978-3-031-87208-2 ISBN 978-3-031-87209-9 (eBook)
https://doi.org/10.1007/978-3-031-87209-9

This Springer imprint is published by the registered company Springer Nature Switzerland AG
The registered company address is: Gewerbestrasse 11, 6330 Cham, Switzerland

In the name of Allah, the Most Gracious and the Most Merciful. First of all, I would like to raise my heartfelt gratitude and appreciation to Allah SWT for the permission, guidance, wisdom, and blessing for all these years till now, when I have reached this important destination of my journey in life to accomplish my goal. Finally, I would like to present my most heartfelt and warmest appreciation to the great parents and parents-in-law (may ALLAH SWT bless and reward them).

Dr. Abdul Sattar Jatoi

Preface

The ever-growing demand for sustainable food production and bioprocessing technologies has transformed the food and bio-industries into dynamic fields of innovation. The integration of food processing with advanced bioprocessing techniques not only enhances productivity and efficiency but also addresses critical challenges such as food security, resource utilization, and environmental sustainability. This book, *Integrated Food and Bioprocessing*, aims to bridge the gap between traditional food processing methods and emerging bioprocessing technologies, offering a comprehensive resource for professionals, researchers, and students alike.

This book covers a wide spectrum of topics, beginning with the principles of food engineering and bioprocessing. It explores the role of biotechnology in food systems, addressing topics such as enzyme applications, fermentation, microbial processing, and the use of novel bioreactors. The text delves into the latest advancements in food preservation, quality control, and product development, emphasizing the importance of interdisciplinary approaches in solving complex food and bioprocessing challenges.

Special attention is given to sustainable practices, including waste valorization, energy-efficient processes, and circular bioeconomy principles. Case studies and real-world examples are incorporated throughout to illustrate how integrated food and bioprocessing technologies are being applied to develop innovative solutions for a range of industries.

This book is designed to serve as both a practical guide and an academic reference, appealing to a broad audience including food scientists, biotechnologists, process engineers, and policymakers. By bringing together diverse perspectives and expertise, it aims to foster collaboration and innovation in these vital sectors.

As the world continues to face pressing challenges related to food security, climate change, and resource scarcity, the integration of food and bioprocessing technologies holds immense promise for shaping a sustainable future. It is our hope that this book will contribute to advancing knowledge, sparking creativity, and inspiring action toward achieving these critical goals.

Karachi, Pakistan — Abdul Sattar Jatoi

Acknowledgments

I thank the Vice Chancellor Engr. Prof. Dr. Samreen Hussain (S.I, T.I) of Dawood University of Engineering and Technology Karachi, and colleagues in the Department of Chemical Engineering for their support and encouragement.

Abdul Sattar Jatoi

Contents

About the Editor

Abdul Sattar Jatoi is an Assistant Professor in the Department of Chemical Engineering at Dawood University of Engineering and Technology, Karachi, Pakistan. He has over 13 years of experience in academics and research. Dr. Jatoi has expertise in food bioprocess and their application in various food industries. Microbial electrochemical systems for environmental and energy applications. He has published over 150 papers, including 50 book chapters and two book. He is an editorial board member of three SCI indexed journal and guest editor of more than three research journals and also peer-reviewed member of more than 50 scientific journals. He has published 70 works in various international conference proceedings. His research interests include: bio-processing, microbial electrochemical system; emerging pollutant treatment via different approaches; sustainable nanomaterials development; wastewater treatment; green-water treatment technology; separation and purification techniques; enviro-analytical techniques.

Part I
Foundations of Integrated Food and Bioprocessing

Chapter 1
History and Importance of Food Processing

Abdul Sattar Jatoi, Arshad Iqbal, Bazil Owais Nasir, Owais Baig, and Saim Akram

1 Introduction

Food processing is the process of converting raw materials into consumable food products using a range of physical, chemical, and biological methods(Knorr & Watzke, 2019). It encompasses a wide range of tasks, from simple ones like cleaning, peeling, and chopping to more complex ones like pasteurization, fermentation, freezing, and preservation use(Tomaszewska et al., 2018). In the past, food processing was developed in response to the human need to preserve food so that it would be available when resources were limited. Food is now a crucial link in the global food supply chain since modern methods have made it safer, more convenient, and available year-round in a variety of ways (Thielecke et al., 2021). The importance of food processing has grown along with industry and urbanization (Bhattu et al., 2024). Large-scale food production, transportation, and preservation became crucial as more people moved from rural to urban regions. Innovations like canning, pasteurization, and refrigeration made it possible to transport food across long distances without sacrificing its safety or quality. This has had significant social and economic repercussions, as it has facilitated the growth of the global food trade and enhanced food security by reducing spoilage and waste. Modern food processing techniques also enable the fortification of foods with essential vitamins and minerals to address nutritional deficiencies in different populations.

Food processing has a crucial role in determining consumer preferences and eating habits in addition to its practical benefits. The development of convenience foods, ready-to-eat meals, and snack products has transformed how people prepare and enjoy their meals in keeping with the fast-paced character of modern life. Although processed foods have benefits like longer shelf life and better food safety,

A. S. Jatoi (✉) · A. Iqbal · B. O. Nasir · O. Baig · S. Akram
Department of Chemical Engineering, Dawood University of Engineering and Technology, Karachi, Pakistan
e-mail: abdul.sattar@duet.edu.pk

A. S. Jatoi (ed.), *Integrated Food and Bioprocessing*,
https://doi.org/10.1007/978-3-031-87209-9_1

they have also generated discussions regarding their effects on nutrition and health, especially when it comes to added sugars, salts, and preservatives. As a result, the food processing sector keeps developing, emphasizing more sustainable and healthful solutions that satisfy consumer and environmental expectations.

2 Ancient Beginnings of Food Processing

In order to survive, early humans developed basic techniques for storing and preserving food throughout the prehistoric era, which is where food processing got its start. Early methods including drying, salting, and fermenting were used to extend the shelf life of food and prevent spoilage during times of scarcity, such as cold or drought. One of the oldest methods was sun-drying cereals, meats, and fruits, which helped to lower moisture content and inhibit the formation of bacteria. A common practice among ancient civilizations, salt, which is easily obtained from the sea, was also an essential preservative for curing meats and fish. The methods used to process food changed along with human societies. Bread, beer, and wine—the mainstays of the ancient Egyptian diet—were produced as a result of their mastery of fermentation. It was well known that the Chinese had invented techniques for fermenting soybeans, which led to the creation of miso, soy sauce, and other seasonings. Large-scale production of alcoholic beverages was made feasible by Mesopotamia's discovery of advanced early brewing technology. Since these advancements in fermentation enhanced food's flavor and nutritional value while also preserving it, they represented a crucial turning point in the history of food processing. Food processing has advanced to a more complex level by the Roman Empire. Romans preserved meat and fish using methods like smoking, which gave them a unique flavor and extended shelf life. Many cultures adopted the practice of pickling vegetables in vinegar and brine, which allowed them to be preserved for months. Furthermore, the increasing significance of processed foods in everyday meals and long-distance trade led to inventions like the production of cheese and the preservation of fruits in sugar and honey. By meeting basic human needs for distribution, storage, and preservation, these early processing techniques established the foundation for the contemporary food industry (Fig. 1.1).

2.1 Middle Ages and Traditional Processing

Throughout the Middle Ages, food processing methods continued to develop, mostly because of trade, necessity, and cultural influences. Food preservation became essential as feudal civilizations expanded and fresh vegetables became limited in many regions. Salting and curing were widely used during this time, particularly for meats like fish and pork. Large quantities of salt were used to preserve these proteins, making them available year-round, especially in areas where refrigeration was not feasible. As a continuation of the curing process, smoking gained popularity as a method of prolonging the flavor and shelf life of meat and fish. Dairy product preservation was another important area of focus during this time. During the Middle Ages, the

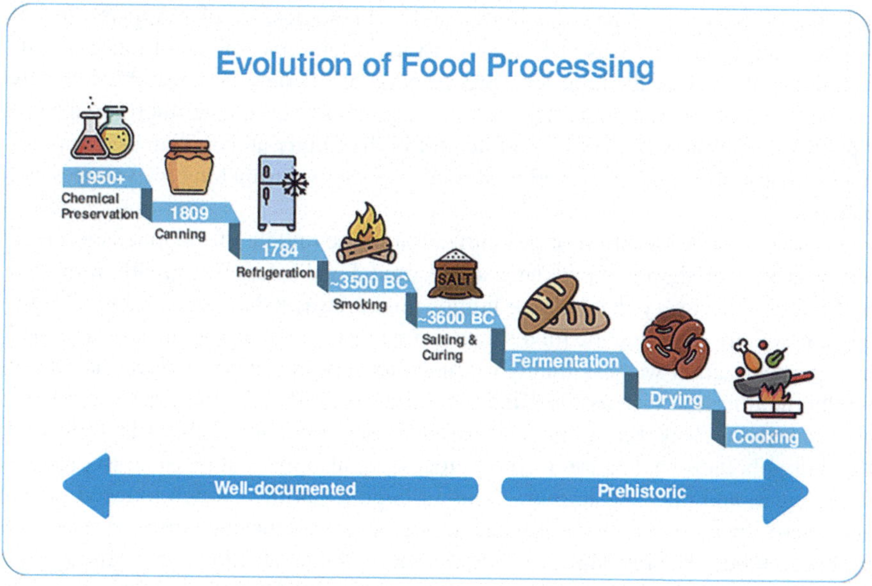

Fig. 1.1 Food processing. (Fardet, 2018)

traditional art of manufacturing cheese was further developed. Monasteries all around Europe developed became hubs for the manufacture of cheese, producing types that were both tradeable and long-storable. In a similar vein, making butter became essential for preserving milk fat, particularly in colder regions. During this time, fermentation flourished as well, and the production of bread and beer continued to develop as vital procedures. New fermentation methods were created, especially for the production of wine, and the use of yeast in baking and brewing improved. During the Middle Ages, spices' use in food processing also increased. It was commonly believed that spices had medicinal and preservation properties in addition to improving food flavor. The expansion of trade routes between Europe and the East, which was made possible by the need for spices like pepper, cinnamon, and cloves, had an additional effect on food preservation methods. Fruits were frequently preserved in sugar or honey, but vegetables were increasingly pickled using vinegar and brine. These methods were especially helpful in controlling seasonal variations in the food supply, giving people access to winter fruits and vegetables that had been preserved. When combined, these antiquated practices influenced medieval diets and established the groundwork for later generations' more sophisticated food processing procedures.

3 Industrial Revolution and Modernization of Food Processing

An important turning point in the history of food processing occurred during the Industrial Revolution, which lasted from the late 18th to the early nineteenth century. Mass production replaced small-scale, artisanal techniques of food production with the introduction of new machinery and technologies. Nicolas Appert's invention of canning in 1809 was one of the most important innovations of this era. This technique allowed for the long-term preservation of a variety of perishable items by sealing them in airtight containers and heating them to eliminate microorganisms. Appert's invention was quickly modified for both military and civilian use, improving the distribution and preservation of food, especially during long sea voyages and battles.

Louis Pasteur's invention of pasteurization in the middle of the nineteenth century was another significant advance at this time. Food safety was greatly impacted by Pasteur's discovery that heating liquids might destroy dangerous bacteria without appreciably changing the food's flavor or quality. Originally applied to wine and beer, pasteurization was eventually extended to milk in order to reduce the risk of foodborne illnesses such as TB. This innovation not only increased food safety but also paved the way for more heat-based preservation methods. The Industrial Revolution's industrialization of food processing also made it possible to produce flour, sugar, and other basic ingredients in large quantities, which reduced prices and improved accessibility. As urbanization grew, so did the need for processed and packaged foods. The introduction of refrigeration in the late nineteenth century significantly transformed food production by enabling the long-distance preservation of perishable items including meat, dairy, and vegetables. The expansion of global food trade was made possible by refrigerated railroad carriages and ships, which linked urban customers with farmers and producers in far-flung regions. Furthermore, processed goods like biscuits, canned soups, and ready-made meals were produced because of the growth of mechanized factories and were commonplace in homes all over the world. These inventions from the Industrial Revolution revolutionized the way food was cooked, distributed, and consumed, paving the way for the modern food industry.

4 Twentieth Century Innovations

Food processing innovation exploded in the twentieth century as a result of shifting consumer habits and scientific breakthroughs. The creation of frozen food technology was one of the most revolutionary. Compared to conventional freezing techniques, Clarence Birdseye's invention of quick-freezing in 1924 significantly improved the texture, flavor, and nutritional content of food. This invention transformed the distribution and storage of food, enabling the widespread availability of frozen fruits, vegetables, meats, and ready-to-cook meals. With the introduction of household freezers in the middle of the twentieth century, the frozen food sector expanded quickly, providing convenience and lowering food waste by prolonging the shelf life of perishable goods. The creation of food additives and preservatives was another significant area of invention in the twentieth century. Chemical preservatives, such ascorbic acid and sodium benzoate, were added to processed foods to prevent spoiling and increase their shelf life. To improve the consistency, taste, and look of products, food scientists also created artificial colorings, flavorings, and emulsifiers. These developments made it possible to produce packaged goods in large quantities that would last for a long time on store shelves. As people's lives became more hectic, processed convenience meals including canned goods, breakfast cereals, instant soups, and snacks became commonplace in homes all over the world. These food additives made food more affordable and accessible, but they also generated discussions about nutrition and health. During the twentieth century, packaging technology also advanced significantly. Food spoilage and waste were decreased by more effective food storage and transportation made possible by innovations like vacuum sealing, Tetra Pak, and plastic packaging materials. The invention of the microwave oven in the 1960s changed how people prepared food at home and increased demand for quick, ready-to-eat meals. By the end of the twentieth century, the food industry had started producing ultra-processed foods that required no preparation and had a long shelf life. Although these advancements increased the accessibility of affordable, easy foods, they also increased knowledge of the harmful health impacts of processed and ultra-processed foods, which helped consumers in the twenty-first century choose healthier choices (Fig. 1.2).

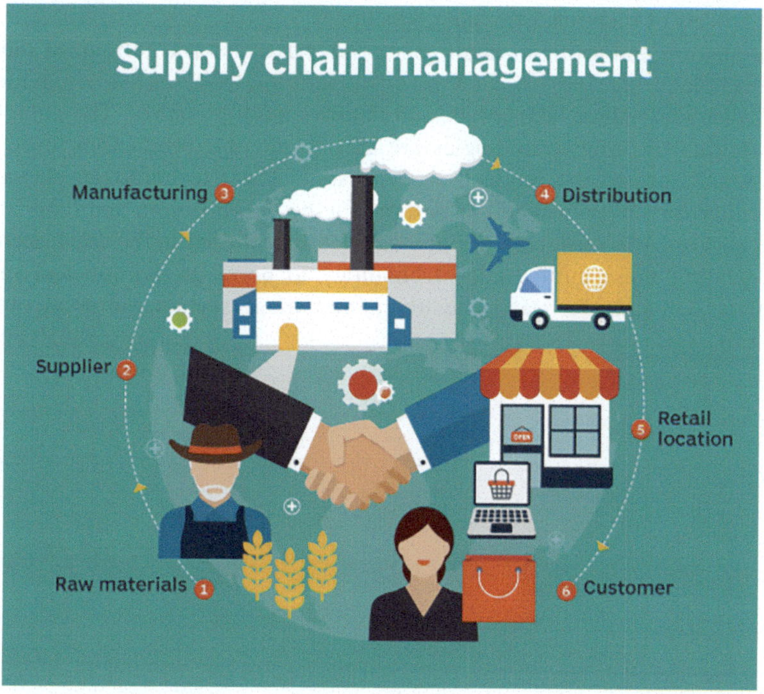

Fig. 1.2 Food processing and supply-chain management. (Dharni & Sharma, 2015)

5 Modern Food Processing Technologies

Significant advancements in food processing technologies have occurred in recent decades because of the need for greater sustainability, efficiency, and food safety. One of the most important developments is High-Pressure Processing (HPP), which uses high-pressure water to kill harmful germs rather than high heat or chemical preservatives. Since HPP extends the shelf life of fresh food while preserving its inherent flavor, color, and nutritional content, it is a popular processing method for products including juices, guacamole, and ready-to-eat meals. This cold pasteurization technique meets consumer demand for minimally processed, "clean label" goods while maintaining safety standards.

Another modern invention is the use of food irradiation, a process that exposes food to controlled levels of ionizing radiation to eradicate pests and pathogens, delay ripening, and extend shelf life. It is commonly used to certain meats, spices, and dried fruits. The public still views irradiation with suspicion despite its remarkable success in ensuring food safety, sometimes due to misconceptions about its effects on food quality. In addition, advanced fermentation methods have gained popularity, particularly for producing protein substitutes. Businesses are now using precision fermentation to create plant-based and lab-grown meats, dairy substitutes, and other sustainable food alternatives in response to concerns about the environmental effects of traditional animal production.

Automation and artificial intelligence (AI) have also revolutionized modern food manufacturing. Large-scale food manufacturing facilities use automation to boost accuracy, reduce labor costs, and improve quality control. These days, AI and machine learning technologies are able to monitor production lines, spot defects, and ensure food products are consistent. These technologies also aid in traceability for safety or recall issues, packaging optimization, and food supply chain tracking. As customer preferences continue to shift, the food sector is implementing these cutting-edge technology to boost production and satisfy consumer demands for safer, healthier, and more sustainable food options (Fig. 1.3).

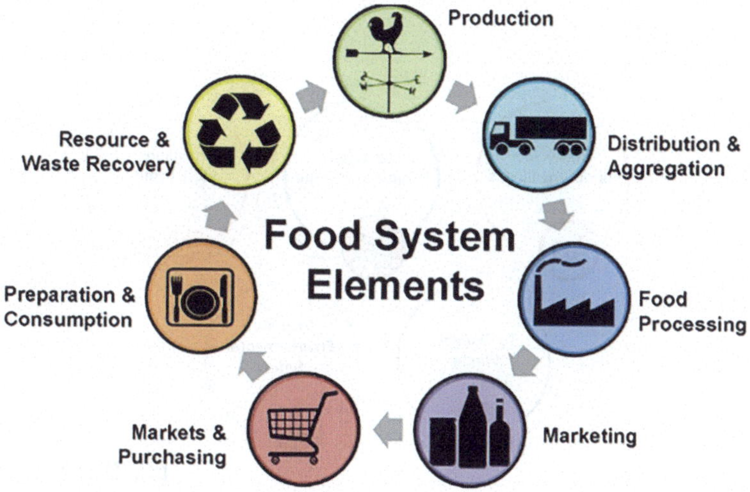

Fig. 1.3 Food chain management system. (Blengini & Busto, 2009)

6 Importance of Food Processing

Food processing is crucial to preserving food safety since it reduces the likelihood of contamination and spoilage. Many raw food products, especially meat, dairy, and fresh fruit, are susceptible to bacterial growth and pathogen contamination if they are not properly preparedFood processing methods, such as pasteurization, sterilization, and high-pressure processing, effectively eliminate harmful bacteria, extending shelf life and reducing the incidence of foodborne illnesses. Customers are reassured about the safety of their food supply, which is important in an increasingly globalized food system where products are sourced and delivered from all over the world, in addition to protecting the health of the public.

In addition to enhancing food safety, food processing significantly increases food accessibility and convenience. It enables the production of a wide variety of easily transportable, storeable, and consumable food items in big quantities. The limitations of seasonality can be overcome by preserving and making year-round available fruits, vegetables, and meats that would otherwise spoil quickly. Processed foods provide consumers with quick and easy meal options that meet their busy schedules. Examples of these include canned items, packaged snacks, and frozen entrees. Additionally, food fortification during processing—for instance, adding vitamins and minerals to wheat or dairy products—helps address nutritional deficiencies in regions where access to fresh, nutrient-dense food may be limited.

Fig. 1.4 Food safety is important for restaurant business. (Ghiselli, 2014)

Food processing is economically vital to the global food supply chain and makes a substantial contribution to the reduction of food waste. Processing techniques that increase the shelf life of perishable items and help prevent spoilage can help preserve and distribute food more efficiently. This is crucial for solving the global issue of food waste, which has consequences for the environment, economy, and ethics. Additionally, by generating millions of employment worldwide, the food processing industry boosts the economies of both rich and developing nations. As the demand for packaged and processed foods increases, food processing remains essential to maintaining a consistent, secure, and sustainable global food supply (Fig. 1.4).

6.1 Challenges in Food Processing

One of the main problems facing the food processing business today is the concern over the health implications of processed and ultra-processed meals. Despite the fact that processing increases food safety and convenience, many processed foods are high in sodium, unhealthy fats, added sugars, and artificial additives. These elements are commonly linked to conditions like obesity, diabetes, and heart disease. Ultra-processed food consumption has been associated with poor dietary quality and an increased risk of chronic illnesses. Ultra-processed foods are often high in calories and low in essential nutrients. As a result, there is growing pressure on the food industry to redesign its products with healthier ingredients and to offer options that would appeal to consumers who are growing more health conscious.

Sustainability and environmental concerns also pose significant challenges to modern food processing. Many traditional processing techniques are resource-intensive because they consume large amounts of water, energy, and packaging materials. The plastic packaging used in processed foods has become a major environmental issue, especially as it adds to plastic waste and ocean contamination. Additionally, food processing facilities generate a lot of waste because of by-products, unused parts, and food deterioration. To address these sustainability issues, the sector must embrace more sustainable sourcing and production practices, develop energy-efficient technology, and provide eco-friendly packaging options.

Ethics are another new issue in food processing, particularly with regard to labor practices, animal welfare, and openness. Around the world, workers in food processing facilities frequently face unsafe working conditions, low pay, and poor working conditions, which has led to calls for stricter labor laws. Awareness of factory farming practices has also increased due to worries about the treatment of animals used in the production of meat and dairy products. Consumers want food companies to disclose more information about the origins of their products, their manufacturing processes, and their effects on the environment and society. To address these ethical concerns, businesses must strike a balance between profit and the need for social and environmental sustainability, which necessitates more transparent and accountable supply chain procedures..

7 Future Trends in Food Processing

One of the most significant upcoming trends in food processing is the shift to sustainable and environmentally friendly practices. As environmental concerns increase, customers and regulatory bodies are putting pressure on food companies to reduce waste, save resources, and reduce their carbon footprint. As a result, less plastic is being used, and more eco-friendly packaging choices are being created, like packaging made of compostable or biodegradable materials. Companies are also trying to find ways to produce food with less water and energy. In the years to come, technological advancements like zero-waste production systems and factories powered by renewable energy sources will likely play a significant role in creating a more sustainable food processing industry.

Consumer demand for "clean label," minimally processed, and healthier products is another significant trend in the food processing industry. As consumers grow more health conscious, they are searching for products free of artificial additives, preservatives, and excessive amounts of sugar or salt. Food producers are therefore revising their recipes to highlight natural ingredients and enhance nutritional value. Consumer desires for nutrient-dense, fresh products can be satisfied by preserving foods using techniques like cold pasteurization and high-pressure processing (HPP) without compromising their natural qualities. The rise of plant-based and alternative proteins, such as lab-grown meat and dairy substitutes, is also transforming the sector by offering sustainable, healthier solutions that meet ethical and nutritional standards.

Thanks to technological advancements, particularly in the fields of automation, biotechnology, and artificial intelligence (AI), food processing is poised for a revolution. AI and machine learning are being used in processing lines to boost efficiency, enhance quality control, and improve food safety. These technologies ensure more homogeneity in the final products and reduce human mistake by enabling real-time process monitoring. Two examples of biotechnology that is opening up new possibilities for the creation of functional meals that are tailored to specific health needs or preferences, like personalized nutrition, are precision fermentation and gene editing. As these technologies improve, they will not only increase production and reduce costs, but they will also enable the creation of innovative food items that meet the demands of a growing global population that is more concerned about their health and the environment.

8 Conclusion

Food processing has played a crucial role in human history, evolving from basic preservation techniques to advanced industrial processes that ensure food security, safety, and convenience. It has enabled mass production, extended shelf life, and improved food availability worldwide. While modern food processing offers numerous benefits, including enhanced nutrition and reduced food waste, concerns regarding additives, over-processing, and environmental impact must be addressed. The future of food processing lies in balancing innovation with sustainability, ensuring that food remains safe, nutritious, and accessible while minimizing negative health and ecological effects.

References

Bhattu, M., Kaur, N., Soni, S. K., & Verma, M. (2024). Factors influencing food processing. In *Nanobiotechnology for food processing and packaging* (pp. 157–172). Elsevier.

Blengini, G. A., & Busto, M. (2009). The life cycle of rice: LCA of alternative agri-food chain management systems in Vercelli (Italy). *Journal of Environmental Management, 90*(3), 1512–1522.

Dharni, K., & Sharma, R. K. (2015). Supply chain management in food processing sector: Experience from India. *International Journal of Logistics Systems and Management, 21*(1), 115–132.

Fardet, A. (2018). Characterization of the degree of food processing in relation with its health potential and effects. *Advances in Food and Nutrition Research, 85*, 79–129.

Ghiselli, R. (2014). Importance of food safety in restaurants. In *Food safety: Researching the hazard in hazardous foods* (pp. 57–80).

Knorr, D., & Watzke, H. (2019). Food processing at a crossroad. *Frontiers in Nutrition, 6*, 85.

Thielecke, F., Lecerf, J.-M., & Nugent, A. P. (2021). Processing in the food chain: Do cereals have to be processed to add value to the human diet? *Nutrition Research Reviews, 34*(2), 159–173.

Tomaszewska, J., Bieliński, D., Binczarski, M., Berlowska, J., Dziugan, P., Piotrowski, J., Stanishevsky, A., & Witońska, I. (2018). Products of sugar beet processing as raw materials for chemicals and biodegradable polymers. *RSC Advances, 8*(6), 3161–3177.

Chapter 2
Bioprocessing

Ibrahim Maina Idriss, Dauda Baba, Faheem Kareem, Abubakar Shetti Kolo, and Zubair Hashmi

1 Introduction to Bioprocessing

Food security, amidst the burgeoning global population, stands as a paramount goal within the principle of sustainable development. The incessant pressures to meet the escalating demands for food production have catalyzed the exploration of various techniques, ranging from the refinement of traditional methods to the implementation of innovative alternatives, including industrial production within controlled laboratory environments. The core imperative remains steadfast: to provide a secure and sustainable supply of food to meet the nutritional needs and health requirements of the global populace. In addressing the concern for the necessity of food security, it is essential to examine the methods of food production and their implications for nutrition and health. Beyond mere sustenance, food serves as a vital source of essential nutrients that underpin human well-being. Consequently, any approach to food production must not only ensure an ample food supply but also guarantee the preservation of safety and nutritional components throughout the production process.

I. M. Idriss (✉) · D. Baba · A. S. Kolo
Department of Chemical Engineering, University of Maiduguri, Maiduguri, Nigeria
e-mail: iimainakaina@unimaid.edu.ng; askolo@unimaid.edu.ng

F. Kareem
Faculty of Integrated Technologies, Universiti Brunei Darussalam, Bandar Seri Begawan, Brunei
e-mail: 21h2401@ubd.edu.bn

Z. Hashmi
Department of Chemical Engineering, Dawood University of Engineering and Technology, Karachi, Pakistan
e-mail: zubair.hashmi@duet.edu.pk

A. S. Jatoi (ed.), *Integrated Food and Bioprocessing*,
https://doi.org/10.1007/978-3-031-87209-9_2

The contemporary landscape of food production is indelibly shaped by the relentless march of technological progress across all facets of human life. This inexorable march has ushered in a new era of food processing methods characterized by rapid advancements in efficiency, waste minimization, and an enhanced focus on the preservation of nutritional value. One of the prominent paradigms that has emerged in advancing crucial processing methods is bioprocessing, an integrative approach that melds biological principles with technological concepts and revolving advancements to optimize food production.

Bioprocessing encompasses a spectrum of techniques that leverage biological systems, such as microorganisms and enzymes, to facilitate the production and preservation of food. In the context of food security, this approach holds promise for several reasons. Firstly, bioprocessing techniques offer a level of precision and control that is conducive to the preservation of safety and nutritional integrity. The ability to harness microorganisms and enzymes in controlled environments allows for targeted interventions, minimizing the risk of contamination and ensuring the retention of essential nutrients in the final food product. Moreover, the integration of bioprocessing into food production aligns with the broader objectives of sustainability. Traditional food production methods often face challenges related to resource depletion, environmental degradation, and the generation of waste. Bioprocessing, with its emphasis on using biological entities, can contribute to more sustainable practices. For instance, the utilisation of microorganisms in fermentation processes not only enhances the flavor and nutritional profile of certain foods but also aids in food preservation, reducing the reliance on chemical additives. The role of bioprocessing is not confined to traditional methods; it extends to innovative alternatives and even industrial-scale production from laboratories. The advent of industrial biotechnology has opened new frontiers in the creation of food products. Through the application of bioprocessing, researchers and food technologists can design and optimize production processes in ways that were previously inconceivable. This not only broadens the spectrum of available food products but also enhances the efficiency and scalability of production, addressing the challenges posed by the growing global demand for food. In essence, the integration of bioprocessing into advanced food processing signifies a paradigm shift toward sustainable and efficient food production. It represents a fusion of biological insights with technological prowess, offering solutions to the intricate challenges posed by food security. As technology continues to advance, the role of bioprocessing is poised to expand, contributing to the evolution of a food production landscape that is not only capable of meeting the needs of a growing population but is also characterized by safety, nutritional richness, and environmental sustainability. As we navigate the complexities of the modern world, the integration of bioprocessing into food production stands as a testament to humanity's capacity to innovate and ensure the well-being of present and future generations.

2 Basic Definitions

Food Any edible substance that contains nourishing and nutritive ingredients that is consumed to provide what the body needs for physical support. The ingredients include carbohydrates, essential minerals, fat and oil, protein, and vitamins serving different roles in the body, such as provision of energy, growth, and other body activities.

Bioprocess The word "**bio**" means any living thing, such as an animal or plant, including microorganisms. A **process** refers to a series of operations that transform raw substances into finished products.

Bioprocessing is simply the use of living cells and their components to transform substances into finished goods.

3 Bioprocess Engineering Principles

Bioprocessing engineering principles involve the application of engineering concepts to manipulate biological systems for efficient production. This field focuses on designing and optimizing bioreactors, the heart of bioprocessing, to ensure controlled and scalable microbial or cell cultures. Engineers in bioprocessing consider factors like mixing, aeration, and temperature control to maximize productivity. The integration of automation, monitoring, and control strategies is paramount to achieving reliable and reproducible bioprocesses for industries ranging from pharmaceuticals to biofuels.

Microbial and Cell Culture

In the realm of bioprocessing engineering principles, microbial and cell culture play pivotal roles in the production of diverse bioproducts. Microbial cultures, including bacteria and yeast, or eukaryotic cell cultures, are cultivated in bioreactors with meticulous attention to environmental parameters such as temperature, pH, and nutrient availability. Bioprocessing engineers focus on optimizing growth conditions, media formulations, and aseptic techniques to foster robust and high-yielding cultures. These principles ensure the scalability, reproducibility, and efficiency of bioproduction processes, contributing to the development of pharmaceuticals, biofuels, and other valuable biotechnological products.

Fermentation Process

Fermentation is a biological process that involves the conversion of sugars into other products, such as organic acids, gases, or alcohol, by microorganisms like bacteria, yeast, or fungi. In the context of bioprocessing, fermentation is a key step in the production of various bioproducts, including pharmaceuticals, biofuels, and industrial enzymes. Figure 2.1 illustrates the fermentation process.

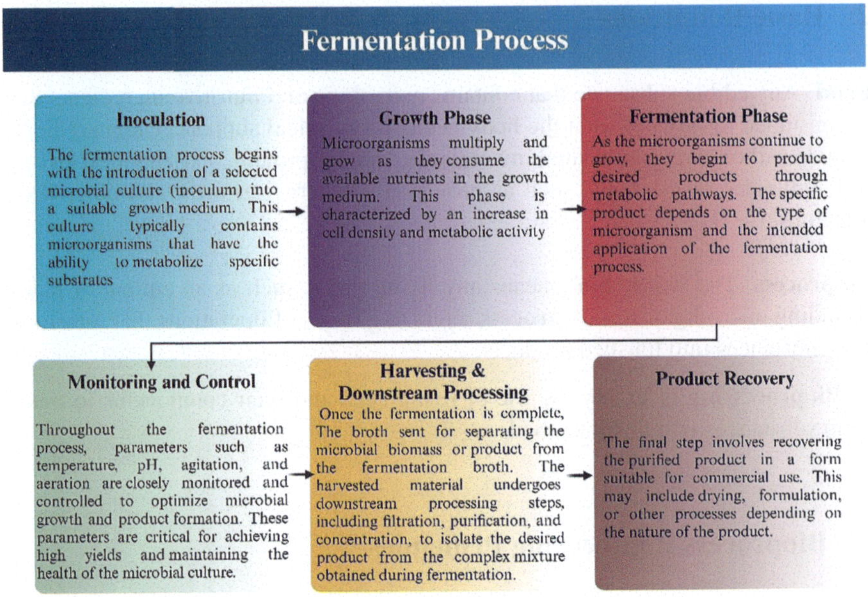

Fig. 2.1 Illuminating the details of the fermentation process in bioprocessing

Fermentation processes are versatile and can be adapted to produce a wide range of products, from antibiotics and vaccines to biofuels and specialty chemicals. The success of a fermentation process relies on a deep understanding of the specific microorganism, the metabolic pathways involved, and the careful control of environmental conditions within the bioreactor.

Downstream Processing

Downstream processing, a pivotal phase in bioprocessing after fermentation or cell culture, plays a crucial role in separating, purifying, and concentrating desired products from complex mixtures generated during upstream processes. This multifaceted procedure involves successive steps to achieve the final, refined product. Initially, microbial biomass or cells are separated from the fermentation broth, typically utilizing centrifugation or filtration techniques. If the product of interest resides within cells, specialized cell disruption methods, such as mechanical homogenization or enzymatic breakdown of cell walls, are applied. Following cell disruption, the mixture undergoes clarification to eliminate solid particles, often through centrifugation or additional filtration.

Chromatographic techniques, including affinity, ion exchange, and size exclusion chromatography, are then employed to selectively separate and purify the target product based on specific properties. The subsequent concentration step involves the removal of excess water, employing techniques like ultrafiltration or precipitation. Sterilization follows to meet stringent quality standards, achieved through filtration or heat treatment. Formulation ensures the stability and compatibility of the final product, often incorporating buffering agents and stabilizers. Ultimately, the processed product is meticulously packaged into its final form, be it vials, bottles, or other containers, with stringent measures maintained to uphold its quality during storage.

Upstream Processing
Upstream processing in bioprocessing encompasses the initial stages of production, involving the cultivation and manipulation of microorganisms or cells to generate the desired bioproduct. The process commences with the introduction of a selected microbial culture or cell line into a growth medium, setting the stage for subsequent stages. Rigorous control of environmental factors, including temperature, pH, aeration, and nutrient availability, fosters the growth and expansion of the microbial population within the bioreactor. Continuous monitoring and adjustment of parameters such as dissolved oxygen, pH, and agitation speed optimize microbial growth and the formation of the desired product, which spans a range of substances from enzymes to pharmaceuticals. Regular sampling ensures the culture's health and the production of the intended product at the expected quality. Harvesting, a critical decision based on growth kinetics and optimal production phases, precedes downstream processing, where microbial contamination is mitigated through sterilization techniques. Depending on the product, recovery from cells may involve methods like cell lysis or filtration. The cultured broth, now containing the product, is then transferred to the downstream processing unit for further separation, purification, and concentration. The efficiency and control exercised in both the upstream processing significantly impact the overall yield, purity, and quality of the final bioproduct, prompting the utilization of advanced technologies and optimization strategies for enhanced productivity in microbial or cell cultures.

3.1 *Bioprocessing Equipment*

Bioprocessing equipment plays a crucial role in the production of bioproducts through the cultivation and manipulation of microorganisms or cells. Here are some key types of bioprocessing equipment commonly used in the industry:

Bioreactors
Bioreactors, essential in bioprocessing, create a controlled environment for microbial or cell culture. Equipped with stirring mechanisms for efficient mixing and aeration, these vessels integrate sensors to monitor and control vital parameters such as temperature, pH, and dissolved oxygen. Airlift Bioreactors represent an alternative type that relies on air bubbles to circulate the culture medium, fostering aeration and mixing without requiring mechanical stirring. Packed Bed Bioreactors, another variant, utilize solid support matrices to immobilize cells or microorganisms, providing a structured environment for growth and product formation. Particularly noteworthy are Single-Use Bioreactors, offering a disposable system designed for convenience. These systems mitigate contamination risks and simplify

cleaning processes, proving especially advantageous in pharmaceutical and biopharmaceutical production.

Fermenters

Batch fermenters represent the simplest form of fermentation, wherein all components are introduced at the outset. This method is well-suited for small-scale productions or processes with a predetermined timeframe. In contrast, continuous fermenters operate by consistently infusing nutrients and removing products, facilitating extended fermentation and maintaining a steady-state operation. Fed-batch fermenters, on the other hand, introduce nutrients incrementally throughout the fermentation process. This approach enhances control over growth phases, optimizing product yield through strategic nutrient management.

Harvesters

Following fermentation, centrifuges play a pivotal role in separating cells or microbial biomass from the fermentation broth, leveraging differences in density. When the target content resides within the cell, high-speed centrifugation is employed for cell disruption, rupturing cells to release intracellular products. Additionally, microfiltration and ultrafiltration employ membranes with specific pore sizes to both clarify and concentrate the fermentation broth. Membrane configuration such as tangential flow filtration (TFF) proves efficient for separating and concentrating high molecular weight substances while minimizing clogging. Chromatography columns are another option, utilizing specific binding interactions between a ligand and the target molecule for separation and purification. This method also separates molecules based on their charge. Moreover, size exclusion chromatography further refines the process by separating molecules based on size, allowing smaller molecules to enter pores in the column while larger ones elute first.

Other Equipment in Bioprocessing

Incubators play a pivotal role in maintaining optimal conditions for microbial or cell culture growth by providing a controlled environment with a stable temperature. Shakers and agitators are instrumental in mixing and homogenization, ensuring the uniform distribution of nutrients, preventing cell sedimentation, and facilitating oxygen transfer in stirred tank bioreactors. Aeration systems aid in the exchange of gases, particularly oxygen, vital for aerobic cultures. Autoclaves contribute to sterilization by using high-pressure steam to sterilize equipment, media, and other items before they are processed. Homogenizers, whether mechanical or ultrasonic, disrupt cells, releasing intracellular products for downstream processing. Analytical instruments, including pH meters and dissolved oxygen sensors, enable real-time monitoring and control of key parameters, ensuring optimal conditions for growth and product formation in highly parametric-dependent bioprocessing. They are predominantly located within the quality control department of the bioprocessing factory. Additionally, freeze dryers (lyophilizers) are employed for preservation and storage).

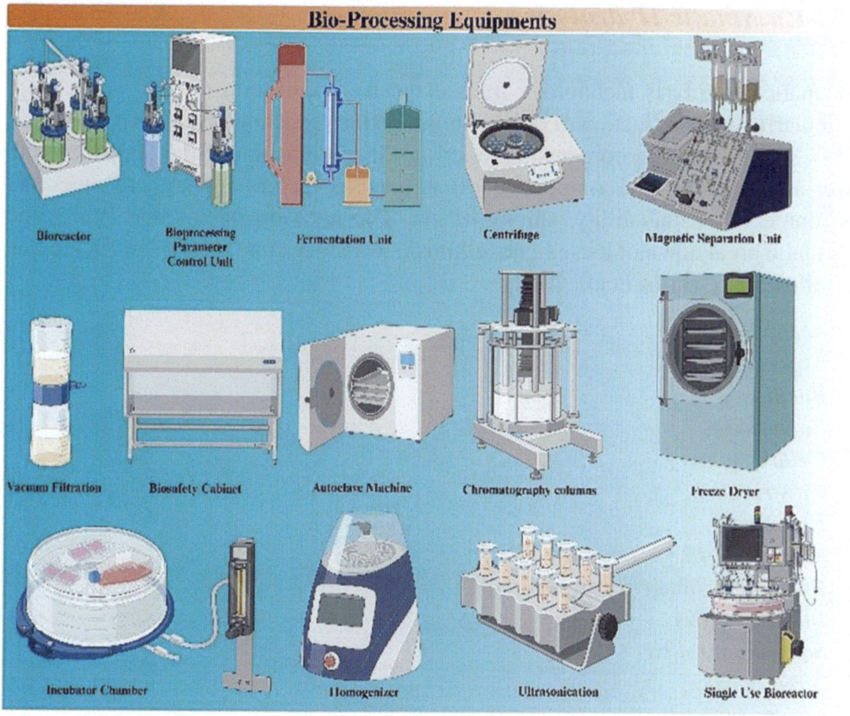

Fig. 2.2 Equipment used during bio-processing

Figure 2.2 shows some of these equipment. Each piece of bioprocessing equipment serves a specific function, collectively contributing to the success and efficiency of bioprocessing operations. The selection of equipment is contingent upon the characteristics of microorganisms, the type of bioproduct, and the specific requirements of the bioprocess.

4 Applications of Bioprocesses in the Food Industry

Bioprocesses find extensive application across diverse food processing industries for the manufacture/or application of materials such as enzymes, flavours, and bakery products. The use of the process for product manufacture helps to maintain the natural characteristics of the product, is less costly, and ensures sustainability among others. Bioprocesses are widely applied in the production of several food ingredients or processed foods.

4.1 *Enzymatic Hydrolysis*

Enzymatic hydrolysis technology finds extensive application in food processing, particularly in the extraction of polyphenols and the reduction of sugars (Qian et al., 2023). The use of microorganisms to aid in the degradation of the feedstock into lower molecular weight components offers benefits such as cost-effectiveness and environmental sustainability, among others. The microorganisms that enable the enzymatic breakdown of these lignocellulosic feedstocks into simple sugars include the following (Canilha et al., 2012):

1. *Acetovibrio*
2. *Aspergillus*
3. *Bacillus*
4. *Bacteriodes*
5. *Cellulomonas*
6. *Clostridium*
7. *Erwinia*
8. *Microbispora*
9. *Phanenerochate chrysosporium*
10. *Ruminococcus*
11. *Sclerotium rolfsii*
12. *Streptomyces*
13. *Thermomonospora*
14. *Trichoderma*

Enzymatic hydrolysis plays a crucial role in the food processing industry to produce a variety of products. This versatile technique contributes to the creation of high-quality food items, ranging from flavor enhancement through protein breakdown to obtaining sweeteners via carbohydrate breakdown into simple sugars. The following are a few examples of products obtainable from enzymatic hydrolysis processes.

4.1.1 Sweeteners

Enzymatic breakdown of starch, sucrose, and insulin is employed in the food processing industry to produce various types of sweeteners. This versatile enzymatic process ensures the creation of a diverse range of sweetening agents tailored towards specific preferences to meet customers' satisfaction. Figure 2.3 illustrates the enzymatic hydrolysis of starch into simple sugars, which serve as sweeteners.

Amylases are essential enzymes that break glycosidic bonds in starch during the liquefaction process, resulting in the production of oligosaccharides and dextrin. Amyloglucosidase, also known as glucoamylase further breaks down these products into glucose monomers in a process called Saccharification (Diler et al., 2015). After the stepwise liquefaction and saccharification stages converted the starch into

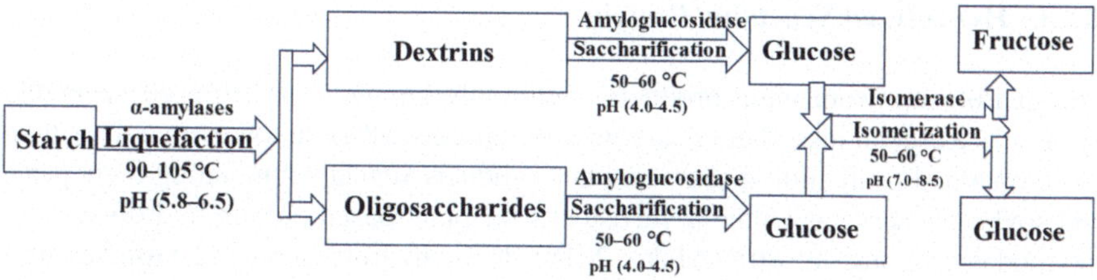

Fig. 2.3 Process of sweetener production

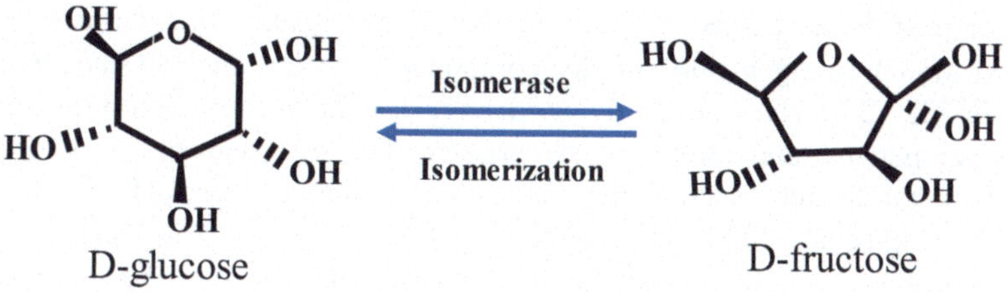

Fig. 2.4 Conversion of glucose to fructose

glucose, the resulting glucose units are further transformed into fructose and smaller glucose units, followed by the purification process. The conversion is accomplished through a process known as isomerization, utilizing an enzyme called isomerase, as depicted in Fig. 2.4. (Nam, 2022).

4.1.2 Hydrolyzed Animal Protein

The utilization of hydrolysis to produce hydrolyzed animal protein (HAP) from various sources, including fish, enhances palatability in meats and baked products. This approach reduces the need for inorganic salts like sodium chloride (Cho et al., 2019). The synthesis of hydrolyzed animal protein (HAP) through enzymatic action on protein- or peptide-containing materials is a crucial process for obtaining organic-based salts, which find application in food processing. It has the advantage of preserving asparagine, glutamine, serine, threonine, and tryptophan, which are either transformed into the corresponding acids partially or wholly get destroyed compared to the use of acid hydrolysis process (Walker & Sweeney, 2002). White bread is an example of food been enhanced using hydrolyzed animal protein (Cho et al., 2019). The application of the naturally occurring salts inform of hydrolyzed animal protein produced with the aid of rightly selected enzyme(s) not only improve the taste of processed foods but also enhances the foods' quality.

4.1.3 Hydrolyzed Vegetable Protein

The hydrolysate from plant products, commonly known as hydrolyzed vegetable protein, is a crucial ingredient that contains nutrients, albeit in lower quantities than those obtained from animal sources. Plant products vegetable including corn, peas, and soybeans, serve as valuable feedstocks for plant-based protein hydrolysates in the food processing industry (Aluko, 2018). These hydrolysates can be used as substitutes for inorganic salts, enhancing the delectability of processed foods. Hydrolyzed vegetable protein is typically produced by washing the protein material to eliminate dirt and contaminants. This is followed by the addition of the appropriate amount of the necessary enzyme(s) to break the bonds in the complex protein material, transforming it into simpler components, protein hydrolysates as illustrated in Fig. 2.5. The commonly used enzymes include α-Chymotrypsin, flavourzyme, papain, and streptomyces griseus. The effectiveness of hydrolysis depends on parameters, such as the enzymes used, temperature, pH, and duration (García Arteaga et al., 2020). The degree of hydrolysis is used to measure the extent of the hydrolysis process in producing both animal and plant-based protein hydrolysates. The degree of hydrolysis (DH) is commonly evaluated using Eq. 2.1 as the proportion of peptide bonds that undergo cleavage during the hydrolytic process (Rutherfurd, 2010), where H_h and H_t represent hydrolyzed and total peptides bonds present and the protein hydrolysate yield in percentage and can be determined using Eq. 2.2 (Islam et al., 2021).

$$\mathrm{DH}(\%) = \frac{H_h}{H_t} \times 100 \tag{2.1}$$

$$\text{Yield} = \frac{\text{Mass of powdered protein hydrolysate}}{\text{Mass of the raw protein material}} \times 100 \tag{2.2}$$

Bond cleavage refers to the fragmentation, rearrangement, or reconstruction of linkages in amino acids or peptides resulting from the hydrolysis of complex protein materials into more soluble amino acids and peptides as illustrated in Fig. 2.6 (Zhang et al., 2015). The breakdown of the protein materials into simpler forms make the amino acids and the peptides more assimilative when ingested.

4.2 *Applications of Fermentation Processes in the Food Industry*

Fermentation is a traditional process inherited from ancient times. It is a natural method that was used in the past to produce alcoholic beverages. Typically, it involved dissolving grain starch in water and allowing it for several days, usually

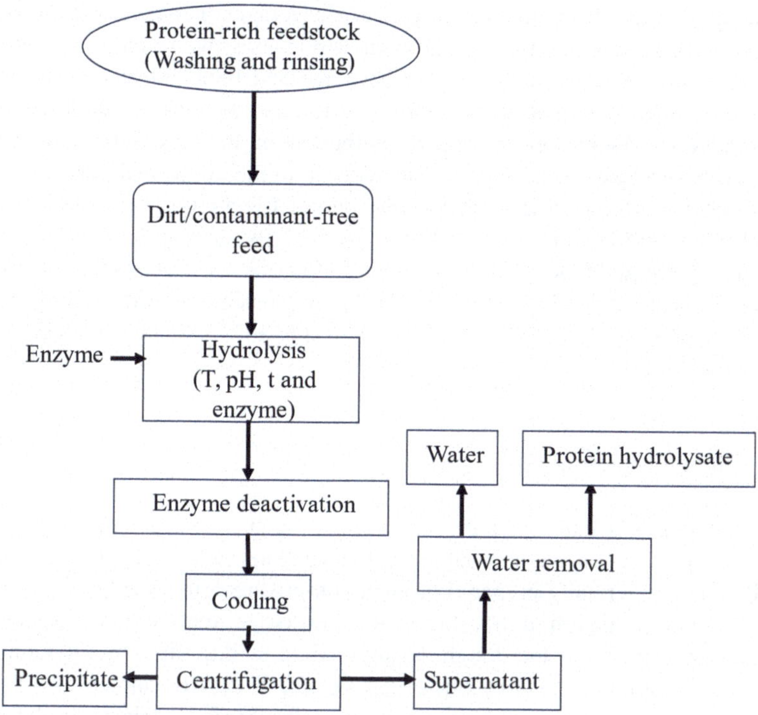

Fig. 2.5 Hydrolyzed vegetable production process

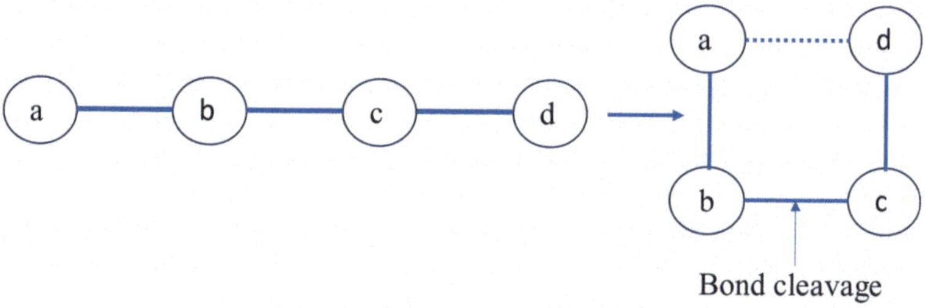

Fig. 2.6 Bond cleavage

about 3 days in an airtight container, to allow microorganisms to act upon it. Fermentation can be defined as the breakdown of organic molecules into simpler ones using microorganisms, especially enzymes called yeast (Martínez-Espinosa, 2020). It is a bioprocess utilized in industries such as food, beverages, bioethanol, pharmaceuticals, and others. The three key components of fermentation are the fermenter, feedstock, and microorganisms acting as the catalyst. The column or vessel in which the fermentation reaction takes place is referred to as the fermenter.

Feedstocks that can be fermented to produce beverages include sugar-containing materials such as maize, sorghum, and corn, and fruits as well as dairy-based feedstocks. The use of fermentation in the beverage and food industry enables food processing, enhances organoleptic characteristics, and extends the shelf life of the food products (Mannaa et al., 2021). Although in the past fermentation was employed to increase these properties in addition to the nutritional benefits, adapting and advancing the process in the conventional technology revealed that fermented foods and beverages offer additional benefits such as less prone to food poisoning, decreased amounts of pathogens and toxins, as well as increased digestion and absorption (Siddiqui et al., 2023). Many foods are produced using the fermentation process, employing microorganisms to facilitate production, as discussed below.

4.2.1 Flour Fermentation

The fermentation of flour is a typical bread-making process that employs enzymes to produce bread. Yeast is commonly used in the baking industry for flour fermentation. It is a commercially prepared product comprising dehydrated cells of one or multiple strains of the fungus Saccharomyces cerevisiae. Yeast serves as a leavening agent, aiding the rise of the dough during baking, as described by the Gompertz equation written represented by Eq. 2.3 (Ali et al., 2012; Romano et al., 2007). In addition to its leavening functions, yeast also contributes to enhancing the flavor and aroma of bread. The breadmaking process begins with the mixing of ingredients, where flour, water, yeast, salt, and other components are combined homogeneously and kneaded into a dough. This mixing stage hydrates the proteins in the flour, leading to the development of gluten—a network of elastic and cohesive molecules, generating carbon dioxide (CO_2) within (Rosentrater & Evers, 2018). The rate of CO_2 production is closely tied to the yeast's metabolism (enzyme), involving the consumption of simple sugars through an irreversible reaction represented in Eq. 2.4

$$y = \alpha \exp\left(\exp\left(\frac{\mu e}{\alpha}\left(t_{\text{lag}} - t\right) + 1\right)\right) \tag{2.3}$$

$$C_6H_{12}O_6 + \text{Yeast} \rightarrow 2CH_3CH_2OH + 2CO_2 + \text{Yeast} \tag{2.4}$$

4.2.2 Meat Fermentation

Meat fermentation has been a practice since ancient times, owing to the presence of beneficial microflora in meats which naturally ferment the meat under observed traditional method. This is typically the practice in the olden days to preserve the meat for future use. One might wonder why meat would spoil prior to the advent of refrigeration systems. The answer is not far-fetched—the groups of flora are found in food materials, including meat and fish into two. While some contribute to beneficial properties such as flavor and aroma, others, characterized as negative microflora overcome the influence of the positive ones, causing meat spoilage (Toldrá, 2007). The process of meat fermentation includes the careful selection of raw meat, the introduction of starter cultures, followed by fermentation, and finally, drying and ripening. Ensuring the quality and hygiene of both the meat and other ingredients is essential for the safety and overall quality of the end product. To initiate and control fermentation, microorganisms like lactic acid bacteria, micrococceae, yeasts, and molds are incorporated into the meat mixture (Toldrá, 2007). The added microorganisms known as lactic acid bacteria (LAB) generate organic acids, primarily lactic acid, by metabolizing sugar, thereby reducing the pH and preventing spoilage and proliferation of pathogenic flora (Kumar et al., 2017). The microorganisms chosen for meat fermentation are selected for based on their capability to produce substances that inhibit spoilage microbes, including bacteriocins, biogenic amines, enzymes, and various other metabolites (Juodeikiene et al., 2012).

The process of meat fermentation involves the selection of quality meat, suitable starter (enzymes), and culture media (salt, carbohydrates, and spices) (Munekata et al., 2022). The mixture is stuffed into casings and subject to controlled temperature and humidity conditions for a certain period. The enzymes such as lactic acid bacteria, Micrococacceae, yeasts, and molds metabolize the carbohydrates in the culture media and generate lactic acid that catalyzes the breakdown of proteins, lipids, and other compounds in the meat, resulting in the formation of volatile and nonvolatile flavor compounds. The lactic acid also lowers the pH and inhibits the growth of spoilage and pathogenic bacteria, thereby extending the product shelf life. The fermented meat is further dried to reduce the water activity at lower temperatures and humidity to enhance the fermented meat product's texture, flavor, and stability. The last stage is the harvesting of the final meat product. Figure 2.7 shows the pictorial representation of the meat fermentation. Lactic acid bacteria (LAB) are the most widely employed microorganisms for meat fermentation, playing various roles in the process. These are tabulated in Table 2.1. Examples of fermented meats consumed around the globe include Sucuk, Salami, Kantwurst, Lup Cheong, Milano Salami, Salami Aeros, Chorizo, Salchichon, and Fuet (Endo et al., 2014).

4.2.3 Fish Fermentation

Fish fermentation process is a technique that uses lactic acid bacteria or other microorganisms to convert fish into more preservable form of protein with improved flavour, taste and other beneficial nutritional compounds. The process of fish fermentation generally commences with fish preparation, involving mixing of the fish with salts to inhibit the growth of spoilage bacteria and fungi. The fermentation includes inoculating the salted fish with enzymes, such as lactic acid bacteria,

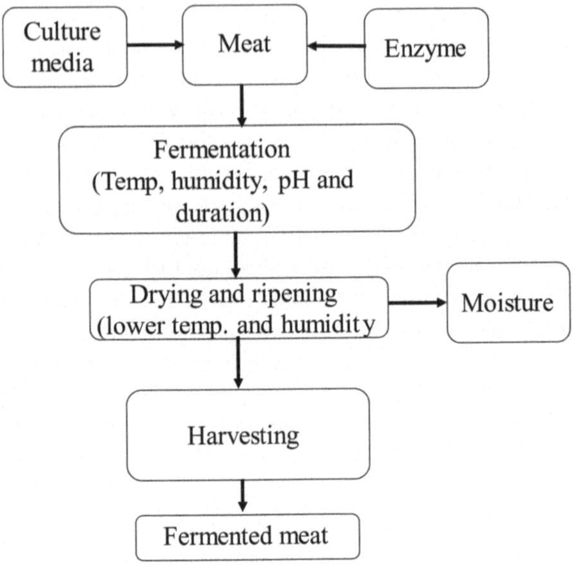

Fig. 2.7 Meat fermentation process

Table 2.1 Role and functions of lactic acid bacteria in meat fermentation (Juodeikiene et al., 2012)

Role	Functions
Starters	Meat fermentation through proteins breakdown, and fatty acid oxidation
Flavor enhancers	Production of low molecular weight peptides, amino acids and volatile compounds from meat proteins bearing desirable flavours.
Bioprotective agents	Production of antimicrobial peptides known as bacteriocins which increase the product shelf live by suppressing the growth and activities of spoilage microorganisms

similar to the meat fermentation process. These added enzymes, called starters, produce organic acids, alcohols, other enzymes, and various metabolites. These compounds then hydrolyze the fish proteins, lipids, and carbohydrates, generating distinct flavor compounds (Marti-Quijal et al., 2020). Examples of starters used in fish fermentation are Bromelain, Chymotrypsin, Papain, Pepsin, Trypsin and factors affecting the fish fermentation process include (Giyatmi & Irianto, 2017):

1. Fish specie and quality

The rate and extent of proteolysis, along with the flavor and aroma of the final product, can be influenced by factors such as the species, size, freshness, and evisceration of the fish. Various fish species possess different quantities and types of endogenous enzymes, playing a crucial role in the initial breakdown of proteins. Eviscerated fish generally exhibit lower enzyme activity compared to whole fish, potentially necessitating a longer fermentation time or the addition of exogenous enzymes.

2. Salt concentration and ratio

Salt introduced to the fish for multiple purposes: to diminish water activity, impede the proliferation of spoilage and pathogenic microorganisms, and improve the flavor and color of the fermented fish. Nevertheless, the impact of salt extends to enzyme activity, affecting both the endogenous fish enzymes and microbial enzymes. Elevated salt concentrations can impede or deactivate these enzymes, leading to a deceleration in proteolysis and a reduced yield of fermented fish. Determining the optimal salt concentration and ratio is contingent upon factors such as the fish type, fermentation method, and the desired quality of the final product.

3. Microbial population and activity

Microorganisms assume a vital role in the fish fermentation process, particularly during the latter stages, where they generate diverse enzymes, organic acids, and volatile compounds, thereby influencing the flavor, aroma, and acidity of the fermented fish. Several factors, including the initial microbial load of the raw fish, salt concentration, pH, temperature, oxygen availability, and the presence of starter cultures or inoculants, affect the microbial population and activity. Skillful selection and management of the microbial population and activity hold the potential to enhance the quality and consistency of fermented fish products.

4. Exogenous enzyme addition

exogenous enzymes may be sourced from plants, animals, or microorganisms, each with distinct specificities, activities, and stabilities. The selection and quantity of exogenous enzymes are contingent upon factors such as the fish type, salt concentration, pH, temperature, and the desired product quality. Additionally, exogenous enzymes can be employed in conjunction with endogenous or microbial enzymes to further augment proteolysis and contribute to the flavor development of the fermented fish. Seafoods including shrimps are fermented for preservation, flavour and nutritional addition. Examples of fermented food are:

1. Fish sauce
2. Bagoong
3. Surströmming
4. Budu
5. Patis

4.2.4 Fermentation of Dairy Products

Raw milk is famous dairy products been processed into a number improved products for consumption in many parts of the world. Milk fermentation is a transformative process wherein bacteria convert lactose into lactic acid. This acid, generated as a byproduct of fermentation, leads to the thickening and souring of the milk, imparting a distinctive tanginess and extending its shelf life. The process involves

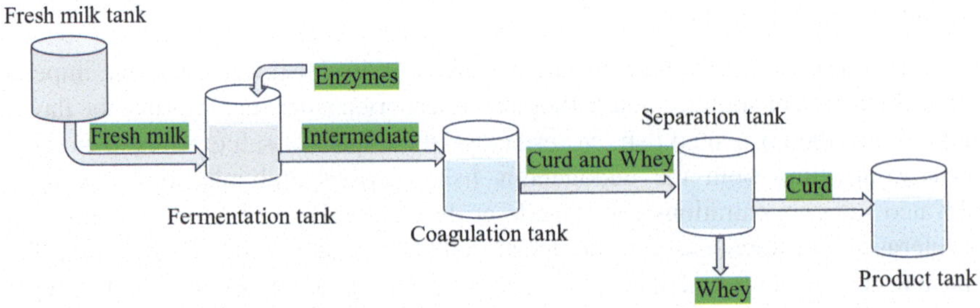

Fig. 2.8 Dairy product fermentation process

inoculating a raw dairy milk with an enzyme, such as lactic acid bacteria, capable of converting lactose to lactic acid. This is followed by incubation under suitable conditions, usually at temperatures between 30 and 40 °C and within a pH range of 4.5–5.5, resulting in the formation of an intermediate product containing a gel-like curd and whey (Kunes et al., 2021). Subsequently, the intermediate is further coagulated to enable the separation of the curd from the whey. The curd is then allowed to ripen into the final desired product, such as yogurt as demonstrated in the Fig. 2.8.

Various factors play a crucial role in determining the quality of the fermented product, including:

1. **The choice of the fresh milk**

Animals such as cows, camels, goats yield milk with varying compositions of fat, protein, lactose, and other nutrients. The variation also transcends to the various species of the same animal type as well as the animal by the milk is obtained. These differences influence the growth and activity of the fermenting enzymes, thereby affecting the curd formation (Hadjimbei et al., 2022).

2. **Fermentation pH**

The **pH level** plays a crucial role in enzymes growth during fermentation of dairy products. Acid conditions favours the flourishing of lactic acid bacteria, which transform lactose to lactic acid and impart flavours and increase the shelf live of the fermentation product. The optimal pH range for milk fermentation is between 4.5 and 5.5 (Zong et al., 2022).

3. **Type of bacteria**

Various bacterial strains contribute distinct flavors and textures through the fermentation process (Bezie & Regasa, 2019). For instance, yogurt is often crafted using Lactobacillus bulgaricus and Streptococcus thermophilus, whereas sour cream is typically produced with Lactobacillus acidophilus (Kunes et al., 2021). Different species of lactic acid bacteria are utilized as starters to initiate the fermentation of milk, resulting in various fermented dairy products. The selection of the starter depends, among other factors, on the type of desired product. For instance, Lactococcus lactis ssp. is employed to catalyze the production of milk products such as butter, buttermilk, and cheese. Lactococcus delbrueckii ssp. bulgaricus catalyze the fermentation of milk into yogurt and probiotic milk. Additionally, kefir can be manufactured by fermenting milk with L. kefir, L. kefiranofacies, and L. brevis (Hati et al., 2013).

4. **Fermentation duration**

The duration of milk fermentation directly impacts the texture and flavor of the end product. Extended fermentation periods yield an enhanced tanginess and a thicker texture (Zong et al., 2022).

5. **Emerging Trent and Future Direction**

In the dynamic landscape of bioprocessing, emerging trends and future directions are shaping the trajectory of sustainable and efficient biomanufacturing. The intersection of technology, biotechnology, and environmental consciousness is fostering groundbreaking approaches. From the rise of single-use bioreactors to the integration of Industry 4.0 principles, these trends reflect a paradigm shift towards precision, sustainability, and advanced process optimization. As bioprocessing evolves, it stands at the forefront of innovation, driven by a commitment to meet global demands while minimizing environmental impact and maximizing efficiency.

- **Single-Use Bioprocessing:**
  Increasing adoption of single-use bioreactors and disposable systems. Enhances flexibility, reduces the risk of contamination, and streamlines cleaning and validation processes.
- **Advanced Bioreactor Design:**
  Continued development of advanced bioreactor designs. Integration of innovative features, automation, and sensors for real-time monitoring to improve control over bioprocess parameters.
- **Data Analytics and Process Optimization:**
  Growing emphasis on data analytics for bioprocess monitoring and optimization. Enables real-time analysis, process adjustments, and predictive modeling to enhance efficiency and yield.
- **Bioprocessing 4.0 and Digitalization:**
  Integration of Industry 4.0 concepts into bioprocessing. Use of digital technologies, artificial intelligence, and automation to create smart, interconnected bioprocessing systems for improved efficiency and decision-making. Industry 4.0 refers to the fourth industrial revolution, encompassing the integration and digitalization of all essential production systems and subsystems, including IoT, AI, cloud computing, robotics, and more (Ilanković et al., 2020).

- **Continuous Bioprocessing:**
  Increasing interest in continuous bioprocessing. Offers potential for higher productivity, reduced footprint, and improved process control compared to traditional batch processes.
- **Cell and Gene Therapy Manufacturing:**
  Growing focus on bioprocessing for cell and gene therapies. Addressing the unique challenges of manufacturing personalized therapies, including scalability, automation, and cost-effectiveness.
- **Synthetic Biology and Metabolic Engineering:**
  Integration of synthetic biology tools for strain development and metabolic engineering. Customizing microorganisms for enhanced production of specific bioproducts, leading to improved yields and process efficiency.
- **Bioproduction of Sustainable and Specialty Chemicals:**
  Exploration of bioprocessing for the production of sustainable chemicals and specialty compounds. Focus on eco-friendly alternatives and the use of renewable feedstocks.
- **Cell-Free Bioprocessing:**
  Increasing interest in cell-free systems for bioprocessing. Offers advantages in terms of flexibility, rapid prototyping, and the ability to produce complex biomolecules without the need for live cells.
- **Sustainability and Green Bioprocessing:**
  Emphasis on sustainable and environmentally friendly bioprocessing. Integration of green technologies, recycling of resources, and reduced environmental impact.
- **Advanced Downstream Processing Techniques:**
  Continued development of advanced downstream processing methods. Improving the efficiency of separation, purification, and concentration processes to meet the demands of complex bioproducts.
- **Global Regulatory Considerations:**
  Evolving regulatory frameworks for bioprocessing. Addressing the challenges associated with the production of advanced therapies and ensuring compliance with quality and safety standards.

It's essential to stay updated with the latest literature, industry publications, and conference proceedings to understand the current trends and future directions in bioprocessing as the field continues to evolve.

5 Conclusion

In conclusion, the field of bioprocessing is dynamic and continually evolving, driven by technological advancements, scientific innovation, and the growing demand for sustainable and efficient production of bioproducts. The integration of cutting-edge bioreactor designs, digitalization, and data analytics has transformed traditional manufacturing processes, enhancing control, flexibility, and overall productivity.

Emerging trends, such as the widespread adoption of single-use bioprocessing systems, continuous bioprocessing, and the application of synthetic biology, reflect a commitment to efficiency, flexibility, and sustainability. The increasing focus on cell and gene therapy manufacturing underscores the industry's response to the evolving landscape of personalized medicine.

As bioprocessing moves towards Industry 4.0 principles, the convergence of digital technologies, artificial intelligence, and automation is reshaping the way bio-production facilities operate. The emphasis on sustainability and green bioprocessing aligns with global efforts to minimize environmental impact and promote eco-friendly alternatives.

Looking ahead, the intersection of bioprocessing with synthetic biology, metabolic engineering, and advanced downstream processing techniques holds promise for unlocking new possibilities in biomanufacturing. Regulatory considerations are also evolving to accommodate the unique challenges posed by novel therapies and ensure the safety and quality of bioproducts. In essence, the future of bioprocessing is marked by a commitment to innovation, sustainability, and the efficient production of diverse bioproducts across industries. Staying abreast of these trends and embracing interdisciplinary approaches will be instrumental in shaping the future landscape of bioprocessing and meeting the demands of a rapidly advancing bioeconomy.

References

Ali, A., Shehzad, A., Khan, M. R., Shabbir, M. A., & Amjid, M. R. (2012). Yeast, its types and role in fermentation during bread making process-a. *Pakistan Journal of Food Sciences, 22*(3), 171–179.

Aluko, R. E. (2018). Food protein-derived peptides: Production, isolation, and purification. In *Proteins in food processing* (pp. 389–412). Elsevier. https://doi.org/10.1016/B978-0-08-100722-8.00016-4

Bezie, A., & Regasa, H. (2019). The role of starter culture and enzymes/rennet for fermented dairy products manufacture- a review. *Nutrition & Food Science International Journal, 9*(2), 0021–0027.

Canilha, L., Chandel, A. K., Dos Santos, S., Milessi, T., Antunes, F. A. F., Da Costa, L., Freitas, W., et al. (2012). Bioconversion of sugarcane biomass into ethanol: An overview about composition, pretreatment methods, detoxification of hydrolysates, enzymatic Saccharification, and ethanol fermentation. *Journal of Biomedicine and Biotechnology, 2012*, 1–15. https://doi.org/10.1155/2012/989572

Cho, Y., Lee, J., Lee, M.-Y., Cho, H.-Y., & Choi, M.-J. (2019). Effects of hydrolyzed animal protein on the enhancement of saltiness and quality characteristics of white pan bread. *Food and Bioprocess Technology, 12*(11), 1832–1841. https://doi.org/10.1007/s11947-019-02332-9

Diler, G., Chevallier, S., Pöhlmann, I., Guyon, C., Guilloux, M., & Le-Bail, A. (2015). Assessment of amyloglucosidase activity during production and storage of laminated pie dough. Impact on raw dough properties and sweetness after baking. *Journal of Cereal Science, 61*, 63–70. https://doi.org/10.1016/j.jcs.2014.10.003

Endo, A., Irisawa, T., Dicks, L., & Tanasupawat, S. (2014). FERMENTED FOODS | fermentations of east and Southeast Asia. In *Encyclopedia of food microbiology* (pp. 846–851). Elsevier. https://doi.org/10.1016/B978-0-12-384730-0.00119-1

García Arteaga, V., Apéstegui Guardia, M., Muranyi, I., Eisner, P., & Schweiggert-Weisz, U. (2020). Effect of enzymatic hydrolysis on molecular weight distribution, techno-functional properties and sensory perception of pea protein isolates. *Innovative Food Science & Emerging Technologies, 65*, 102449. https://doi.org/10.1016/j.ifset.2020.102449

Giyatmi, G., & Irianto, H. E. (2017). Enzymes in fermented fish. In *Advances in food and nutrition research* (Vol. 80, pp. 199–216). Elsevier. https://doi.org/10.1016/bs.afnr.2016.10.004

Hadjimbei, E., Botsaris, G., & Chrysostomou, S. (2022). Beneficial effects of yoghurts and probiotic fermented milks and their functional food potential. *Food, 11*(17), 2691. https://doi.org/10.3390/foods11172691

Hati, S., Mandal, S., & Prajapati, J. (2013). Novel starters for value added fermented dairy products. *Current Research in Nutrition and Food Science Journal, 1*(1), 83–91. https://doi.org/10.12944/CRNFSJ.1.1.09

Ilanković, N., Zelić, A., Gubán, M., & Szabó, L. (2020). Smart factories – The product of Indrusty 4.0. *PRO, 7*(1), 19–31. https://doi.org/10.31570/Prosp_2020_01_2

Islam, M. S., Hongxin, W., Admassu, H., Noman, A., Ma, C., & An Wei, F. (2021). Degree of hydrolysis, functional and antioxidant properties of protein hydrolysates from grass turtle (*Chinemys reevesii*) as influenced by enzymatic hydrolysis conditions. *Food Science & Nutrition, 9*(8), 4031–4047. https://doi.org/10.1002/fsn3.1903

Juodeikiene, G., Bartkiene, E., Viskelis, P., Urbonaviciene, D., Eidukonyte, D., & Bobinas, C. (2012). Fermentation processes using lactic acid bacteria producing bacteriocins for preservation and improving functional properties of food products. In M. Petre (Ed.), *Advances in applied biotechnology*. InTech. https://doi.org/10.5772/30692

Kumar, P., Chatli, M. K., Verma, A. K., Mehta, N., Malav, O. P., Kumar, D., & Sharma, N. (2017). Quality, functionality, and shelf life of fermented meat and meat products: A review. *Critical Reviews in Food Science and Nutrition, 57*(13), 2844–2856. https://doi.org/10.1080/10408398.2015.1074533

Kunes, R., Bartos, P., Iwasaka, G. K., Lang, A., Hankovec, T., Smutny, L., Cerny, P., Poborska, A., Smetana, P., Kriz, P., & Kernerova, N. (2021). In-line technologies for the analysis of important milk parameters during the milking process: A review. *Agriculture, 11*(3), 239. https://doi.org/10.3390/agriculture11030239

Mannaa, M., Han, G., Seo, Y.-S., & Park, I. (2021). Evolution of food fermentation processes and the use of multi-omics in deciphering the roles of the microbiota. *Food, 10*(11), 2861. https://doi.org/10.3390/foods10112861

Martínez-Espinosa, R. M. (2020). Introductory chapter: A brief overview on fermentation and challenges for the next future. In R. M. Martínez-Espinosa (Ed.), *New advances on fermentation processes*. IntechOpen. https://doi.org/10.5772/intechopen.89418

Marti-Quijal, F. J., Remize, F., Meca, G., Ferrer, E., Ruiz, M.-J., & Barba, F. J. (2020). Fermentation in fish and by-products processing: An overview of current research and future prospects. *Current Opinion in Food Science, 31*, 9–16. https://doi.org/10.1016/j.cofs.2019.08.001

Munekata, P. E. S., Pateiro, M., Tomasevic, I., Domínguez, R., Da Silva Barretto, A. C., Santos, E. M., & Lorenzo, J. M. (2022). Functional fermented meat products with probiotics – A review. *Journal of Applied Microbiology, 133*(1), 91–103. https://doi.org/10.1111/jam.15337

Nam, K. H. (2022). Glucose isomerase: Functions, structures, and applications. *Applied Sciences, 12*(1), 428. https://doi.org/10.3390/app12010428

Qian, J., Chen, D., Zhang, Y., Gao, X., Xu, L., Guan, G., & Wang, F. (2023). Ultrasound-assisted enzymatic protein hydrolysis in food processing: Mechanism and parameters. *Food, 12*(21), 4027. https://doi.org/10.3390/foods12214027

Romano, A., Toraldo, G., Cavella, S., & Masi, P. (2007). Description of leavening of bread dough with mathematical modelling. *Journal of Food Engineering, 83*(2), 142–148. https://doi.org/10.1016/j.jfoodeng.2007.02.014

Rosentrater, K. A., & Evers, A. D. (2018). Bread-baking technology. In *Kent's technology of cereals* (pp. 565–622). Elsevier. https://doi.org/10.1016/B978-0-08-100529-3.00008-6

Rutherfurd, S. M. (2010). Methodology for determining degree of hydrolysis of proteins in hydrolysates: A review. *Journal of AOAC International, 93*(5), 1515–1522. https://doi.org/10.1093/jaoac/93.5.1515

Siddiqui, S. A., Erol, Z., Rugji, J., Taşçı, F., Kahraman, H. A., Toppi, V., Musa, L., Di Giacinto, G., Bahmid, N. A., Mehdizadeh, M., & Castro-Muñoz, R. (2023). An overview of fermentation in the food industry – Looking back from a new perspective. *Bioresources and Bioprocessing, 10*(1), 85. https://doi.org/10.1186/s40643-023-00702-y

Toldrá, F. (2007). Fermented meat production. In Y. H. Hui (Ed.), *Handbook of food products manufacturing* (1st ed., pp. 265–279). Wiley. https://doi.org/10.1002/9780470113554.ch60

Walker, J. M., & Sweeney, P. J. (2002). Production of protein hydrolysates using enzymes. In *The protein protocols handbook* (2nd ed., pp. 563–566). Humana Press.

Zhang, D., Zhou, J., Xia, F., Kang, Z., & Hu, W. (2015). Bond cleavage, fragment modification and reassembly in enantioselective three-component reactions. *Nature Communications, 6*(1), 5801. https://doi.org/10.1038/ncomms6801

Zong, L., Lu, M., Wang, W., Wa, Y., Qu, H., Chen, D., Liu, Y., Qian, Y., Ji, Q., & Gu, R. (2022). The quality and flavor changes of different soymilk and milk mixtures fermented products during storage. *Fermentation, 8*(12), 668. https://doi.org/10.3390/fermentation8120668

Chapter 3
Sustainable Food Production

Abdul Sattar Jatoi, Arshad Iqbal, Muhammad Azam Usto, and Bazil Owais Nasir

1 Introduction

Sustainable food production is a farming and food systems approach that aims to meet the nutritional needs of the present population while preserving the resources required for future generations(Smith & Gregory, 2013). Food security, environmental protection, and social justice are all balanced. With an estimated 10 billion people on the earth by 2050, sustainable food production is crucial to addressing problems including environmental degradation, food scarcity, and resource depletion(Bale et al., 2008). Soil erosion, biodiversity loss, and an increase in greenhouse gas emissions are just a few of the major environmental issues brought on by traditional agricultural methods, which usually rely on monoculture and intensive chemical inputs(McKenzie & Williams, 2015). On the other hand, sustainable food production seeks to lessen these negative impacts by putting into practice methods that maintain or improve ecosystem health while producing enough nutritious food. Resource conservation, social responsibility, and reducing environmental effects are the three fundamental principles of sustainable food production(Dobermann & Nelson, 2013). This includes reducing dependency on artificial inputs like fertilizers and pesticides, promoting local food systems to cut down on food miles, conserving water, and improving soil health through crop diversity and organic amendments. Importantly, sustainable agriculture also seeks to boost biodiversity and restore harmed ecosystems(Jatoi & Mubarak, 2024). These strategies help mitigate climate change and strengthen resilience to climate-related risks like droughts, floods, and shifting agricultural seasons by reducing emissions and sequestering carbon(Sattar Jatoi et al., 2024). For instance, agroecology, which integrates ecological principles into farming,

A. S. Jatoi (✉) · A. Iqbal · M. A. Usto · B. O. Nasir
Department of Chemical Engineering, Dawood University of Engineering and Technology, Karachi, Pakistan
e-mail: abdul.sattar@duet.edu.pk

A. S. Jatoi (ed.), *Integrated Food and Bioprocessing*,
https://doi.org/10.1007/978-3-031-87209-9_3

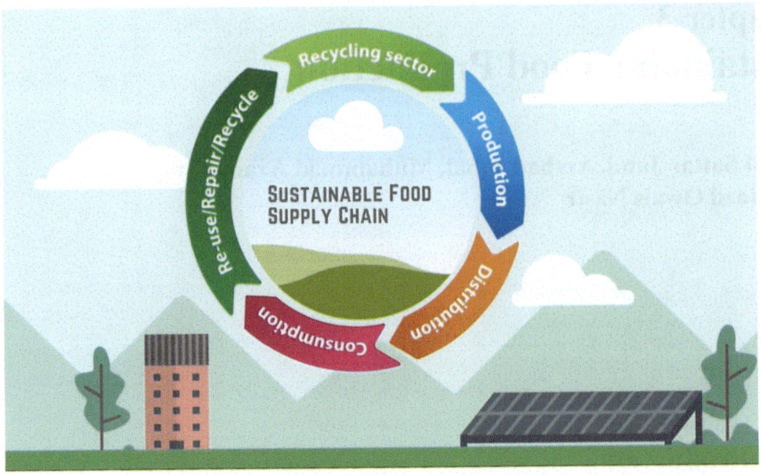

Fig. 3.1 Sustainable food supply chains: meeting demand and reducing waste. (Smith, 2008)

reduces the need for external inputs by promoting the use of natural processes for soil fertility and pest control. Social and economic aspects are equally crucial for sustainable food production. Important objectives include assisting rural communities, ensuring that farmers are compensated, and creating equitable access to resources and markets. Sustainable food systems also consider the welfare of workers and animals and promote fair trade policies(Khalid et al., 2024). Growing consumer expectations for openness and sustainability in food production have led to the expansion of certification schemes (like organic and fair trade) that incentivize farmers to employ more sustainable practices (Smith, 2008). That must be incorporate technology, improve legal frameworks, and foster international cooperation to make food production systems more resilient, sustainable, and capable of feeding the world's growing population while preserving its ecosystems (Hamid et al., 2024) as highlighted in Fig. 3.1.

2 Sustainable Farming Practices

In order to meet the current demands for food production while preserving resource availability and environmental health for future generations, sustainable farming methods were developed (Dessart et al., 2019). Among the most crucial methods is organic farming, which uses natural processes instead of synthetic chemicals like fertilizers and insecticides to increase soil fertility and control pests(Serebrennikov et al., 2020). Because this agricultural approach improves soil health, promotes biodiversity, and reduces pollutants, it is more environmentally friendly. Similarly, crop rotation and variety prevent soil depletion, reduce pest and disease outbreaks, and enhance ecological balance(Meyer-Aurich, 2005). These methods improve soil fertility over time by introducing cover crops that stop soil erosion and boost nutrient levels, as well as rotating crops with different nutrient requirements. Another crucial method is agroecology, which integrates ecological principles into agricultural systems. Agroecology emphasizes biodiversity, natural resource management, and indigenous knowledge to create resilient, self-sustaining ecosystems(Balogun et al., 2022; Meyer-Aurich, 2005). Farmers are encouraged to mimic natural processes by using techniques like integrated pest management (IPM), which uses biological control methods including beneficial insects, natural predators, and crop rotations to manage pest populations while reducing the need of chemical pesticides(Brito et al., 2021). Permaculture, which centers agricultural systems around natural ecosystems and promotes closed-loop systems that recycle waste and efficiently use resources, is another sustainable strategy. These practices strengthen farms' resilience to environmental pressures like drought and climate change by fostering healthy ecosystems and reducing the requirement for external inputs(Muhie, 2022). Advances in technology are also essential to sustainable farming. Precision agriculture allows farmers to optimize resources, reduce waste, and mitigate their adverse environmental impacts by using GPS, drones, and sensors to monitor crop health, water use, and soil conditions(Gamage et al., 2023). Conservation tillage helps retain moisture, reduce erosion, and preserve soil structure by minimizing or doing away with the need to plow the land. Two sustainable livestock practices that enhance animal health and reduce the environmental impact of meat production are rotational grazing and free-range farming(Rust et al., 2021). By combining established practices with cutting-edge technologies, sustainable farming not only boosts productivity but also strengthens resistance to issues brought on by biodiversity loss, resource shortages, and climate change (Fig. 3.2).

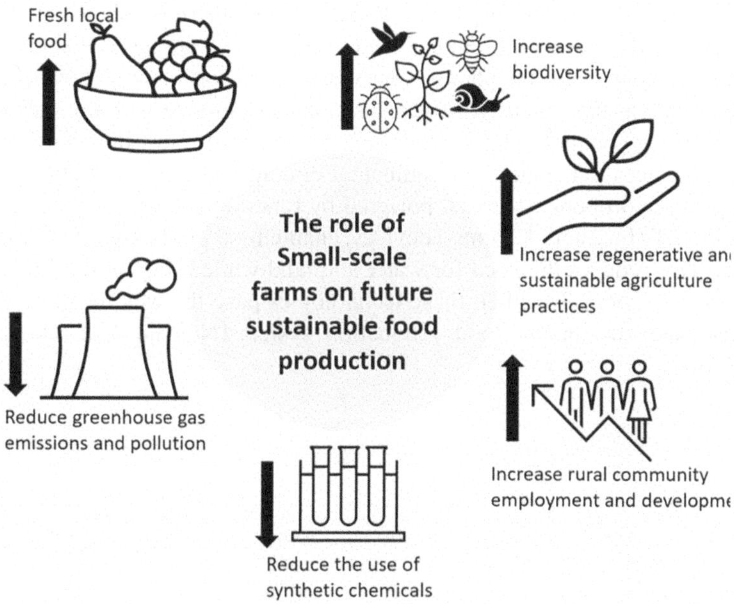

Fig. 3.2 Adaptation of resilient regenerative agricultural practices by small-scale growers towards sustainable food production in north-central Arizona. (Mpanga et al., 2021)

3 Role of Technology in Sustainable Food Production

Technology makes food production far more sustainable by increasing output, reducing waste, and minimizing environmental impacts(Gorjian et al., 2021). Precision agriculture, which uses data from GPS, drones, and satellite photos to monitor crop health, soil moisture, and nutrient levels, is one important technological advancement(Boix-Fayos & de Vente, 2023). This technology reduces environmental harm and the need for needless inputs by empowering farmers to make informed decisions about the use of pesticides, fertilizer, and water. Through clever irrigation systems that only utilize water where it is needed, precision farming can also help preserve water by optimizing water use in drought-prone areas. By increasing productivity while consuming fewer resources, precision agriculture contributes to the development of more robust and sustainable food systems(Chopin et al., 2021). Biotechnology is another important technological development supporting sustainable food production. Developments in genetic engineering and selective breeding reduce the need for chemical inputs and enable cultivation in marginal areas by producing crops that are more resistant to pests, diseases, and extreme weather. Because GMOs have been designed to increase crop output, improve nutritional content, and tolerate drought, they are particularly useful in addressing food security in regions that are vulnerable to climate change. Biotechnology also encompasses microbial innovations such as biopesticides and biofertilizers, which reduce dependence on synthetic chemicals by using natural organisms to preserve crops and promote plant development(Slimi et al., 2021).

Additionally, smart farming technologies like automation, robots, and artificial intelligence (AI) are revolutionizing food production. Automated machinery may do labor-intensive tasks like planting, harvesting, and sorting crops, reducing labor costs and increasing output.AI-driven systems can evaluate vast amounts of agricultural data to predict future trends and optimize farming practices, allowing farmers to react to changing environmental circumstances more quickly. Vertical farming and hydroponic systems, powered by renewable energy sources and controlled by IoT (Internet of Things) devices, enable food production in metropolitan settings. This reduces the need for water and land while bringing food production closer to consumers. Together, these technologies pave the way for more flexible, effective, and sustainable food production systems(Nguyen & Drakou, 2021) (Fig. 3.3).

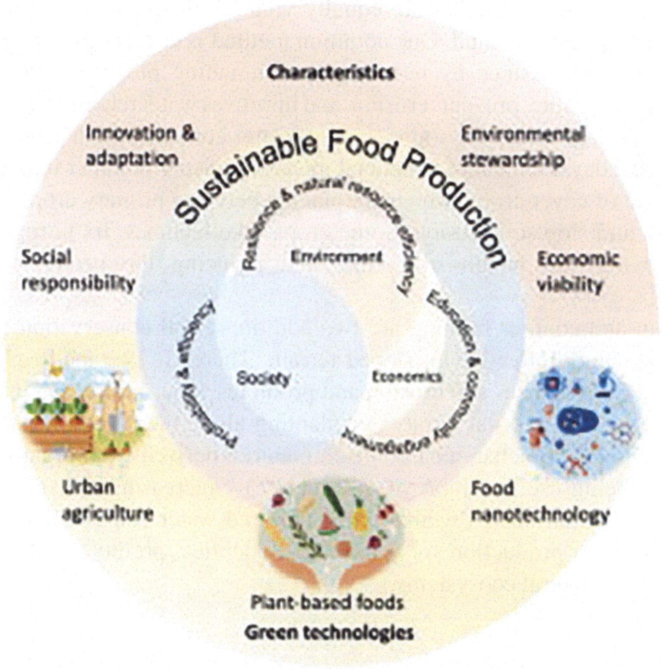

Fig. 3.3 Perspectives on sustainable food production system: characteristics and green technologies. (Qu et al., 2024)

4 Water and Soil Conservation Techniques

Soil and water conservation is essential to sustainable food production because it ensures that these essential resources are preserved and optimized for future agricultural use. Drip irrigation is one of the finest ways to manage water since it uses a network of emitters and tubes to transport water directly to plant roots, reducing water waste from evaporation or runoff. This strategy is extremely beneficial when water is scarce in arid and drought-prone regions(Hoek et al., 2021). Drip irrigation systems can be equipped with moisture sensors, which will only activate when soil moisture levels are low, to further increase water-use efficiency. Another option to conserve water is to use storage systems and rainwater harvesting. By storing rainfall for use during dry spells, these techniques lessen dependency on groundwater and prevent soil erosion caused by intense rainfall(Setsoafia et al., 2022).

Soil conservation techniques are equally vital for maintaining the productivity and health of agricultural land. One common method is conservation tillage, which minimizes soil disturbance by reducing or eliminating plowing. This technique maintains soil structure, prevents erosion, and improves water retention. Additionally, conservation tillage raises the soil's organic matter content, which promotes biodiversity and the development of beneficial microorganisms. Another important strategy is the use of cover crops, which are planted between primary crops to improve soil fertility and stop soil erosion. Some crops, like legumes, fix nitrogen into the soil, improving soil health over time and reducing the need for chemical fertilizers.

Terracing and contour farming are two additional soil conservation techniques that are particularly effective on sloped terrain. There are two methods to reduce water flow, which reduces soil erosion and promotes water infiltration into the soil: contouring the land into flat terraces or planting along the natural contours of the slope. Additionally, biochar, a carbon-rich material derived from organic waste, is increasingly being used to improve soil quality by increasing the soil's ability to retain water and nutrients. Together, these soil and water conservation techniques provide long-term production for future food supplies, promote sustainable land use, and protect natural ecosystems.

5 Sustainable Livestock and Aquaculture Practices

Reducing the adverse environmental effects of meat and dairy production while ensuring economic viability, enhancing biodiversity, and increasing animal welfare are the objectives of sustainable livestock operations. One of the most crucial tactics is rotational grazing, which entails transferring cattle between pasture areas to prevent overgrazing and encourage vegetation regrowth. This method improves the nutritional value of cow feed, reducing the need for external feed inputs, while also preserving healthy grasslands, which are crucial for holding onto water and storing carbon in the soil. In free-range and grass-fed systems, animals' natural behavior is also prioritized, reducing reliance on feedlots, which are often resource-intensive and produce a lot of waste. By improving animal health and land use, these technologies contribute to a reduction in greenhouse gas emissions associated with livestock farming.

In addition to land management practices, sustainable livestock production includes waste minimization and the utilization of alternative feed sources. Methane emissions from ruminants, such as sheep and cows, are a significant contributor to greenhouse gas emissions. To counteract this, researchers are investigating dietary modifications such as the addition of seaweed or specific feed additives that reduce the quantity of methane generated during digestion. Additionally, sustainable livestock systems often use by-products from other industries, such as agricultural wastes or food wastes, as feed to save costs for farmers and lessen the environmental effect of feed production. By implementing better waste management practices, such as anaerobic digestion to convert manure into biogas for energy production, methane emissions are reduced and renewable energy is generated.

Sustainable aquaculture, or fish farming, is another crucial component of sustainable food production, especially as the demand for seafood rises. Water contamination, habitat damage, and antibiotic abuse are all possible outcomes of conventional aquaculture. Integrated multi-trophic aquaculture (IMTA) has emerged as a sustainable approach to these issues by simulating natural ecosystems. By farming several species together in a way that allows waste from one species (like fish) to be used as nutrients by others (like seaweed or shellfish), IMTA lowers pollution and establishes a balanced, resource-efficient system. Recirculating aquaculture systems (RAS), which filter and reuse water, are another sustainable method that significantly reduces water use and waste discharge into the environment. The long-term sustainability of animal-based food production is ensured by these innovative aquaculture and livestock practices, which also help to prevent environmental harm.

6 Food Waste Reduction and Post-harvest Management

Food waste is a significant issue in global food systems since about one-third of all food produced for human consumption is lost or wasted. Because it conserves energy, water, and land and reduces greenhouse gas emissions from landfills, reducing food waste is crucial to achieving sustainable food production. Food waste can occur at any stage of the supply chain, from manufacture to consumption. As soon as crops are harvested, post-harvest management is crucial to reducing losses. This includes improved storage facilities, such as climate-controlled warehouses and silos, which control humidity, temperature, and pest infestations to keep food from rotting. Better handling techniques, including as timely harvesting and efficient transportation, also reduce post-harvest losses, especially in developing countries with possibly insufficient infrastructure.

Innovative preservation techniques are essential to extending the shelf life of perishable items. Using techniques like cold chain logistics, which involves keeping food at optimal temperatures during production, storage, and transportation, is necessary to reduce spoilage in fruits, vegetables, dairy, and meat products. Additionally, utilizing advanced packaging solutions, such as vacuum sealing and biodegradable materials, can help prevent infection and preserve food freshness. Creating value-added products from food, like sauces, jams, or dried goods, is another method to cut down on waste. Food processing, which often turns potential waste into durable and marketable items, allows more of the generated crop to reach customers.

Increasing knowledge and encouraging better buying, storing, and consumption habits are key components of addressing food waste at the consumer level. By educating people about portion management, proper food storage, and the importance of eating "imperfect" fruits and vegetables, household and restaurant waste is reduced. Food recovery and redistribution programs are also crucial; they rescue

Fig. 3.4 Reducing fresh produce losses and increasing efficiency in harvest & post-harvest. (Trimble, 2021)

surplus food from farms, eateries, and retail stores and provide it to those in need. These initiatives address hunger and food waste at the same time by linking excess food with food banks and community kitchens. In order to ensure that the food produced is consumed rather than wasted, post-harvest management techniques and food waste reduction are crucial for developing more effective and sustainable food systems(Candemir et al., 2021) (Fig. 3.4).

7 Policies and Global Initiatives Supporting Sustainable Food Production

These initiatives fight hunger and food waste at the same time by linking excess food with food banks and community kitchens. Because they ensure that the food produced is consumed rather than wasted, post-harvest management techniques and food waste reduction are crucial for developing more effective and sustainable food systems(Short et al., 2021). International Programs and Regulations to Encourage Sustainable Food Production. Strong worldwide actions and policies are needed to address the social, economic, and environmental challenges facing agriculture and to advance sustainable food production(Leeuwis et al., 2021). A significant global framework, the United Nations Sustainable Development Goals (SDGs), especially Goal 2: Zero Hunger, seek to end hunger, attain food security, enhance nutrition, and promote sustainable agriculture. This goal encourages countries to adopt practices that increase agricultural productivity while preserving ecosystem health, particularly in regions vulnerable to resource scarcity and climate change. Fair resource access, sustainable land and water management, and support for smallholder farmers—who produce a significant amount of the world's food but often lack the resources to practice sustainable agriculture—are all emphasized in the SDGs.

Different strategies are being implemented by governments worldwide to promote sustainable food production. Many countries have put in place agricultural subsidies and incentives to encourage farmers to adopt sustainable techniques like organic farming, agroforestry, or using less chemicals. As an illustration, consider the European Union's Common Agricultural Policy (CAP), which links significant subsidies to environmentally beneficial farming practices (Fanzo et al., 2021). In a similar vein, programs such as India's Pradhan Mantri Krishi Sinchayee Yojana, which offer subsidized drip irrigation systems, promote efficient water management in developing countries. These policies promote sustainable agricultural methods and rural development by increasing farmers' incomes and quality of life. In addition to national regulations, international organizations and global initiatives are driving change(Ng et al., 2021). The Food and Agriculture Organization (FAO) and the Global Alliance for Climate-Smart Agriculture (GACSA) work together with the public and private sectors to promote sustainable farming practices that reduce greenhouse gas emissions and boost resilience. Certification programs, including Fair Trade, Rainforest Alliance, and Organic Certification, provide market incentives to producers who meet specific environmental and social sustainability standards in an effort to persuade consumers to adopt sustainable food systems(Vittuari et al., 2021). Growing customer demands for sustainability and transparency, which encourage businesses to adopt more ecologically friendly supply chain practices, promote these initiatives. These legislation and international initiatives encourage a collaborative approach to sustainable food production by addressing the interrelated problems of social equity, environmental degradation, and food security.

8 Challenges and Future Trends in Sustainable Food Production

Sustainable food production has a number of challenges, particularly as global population expansion and climate change put more strain on agricultural systems. One of the primary problems is climate change, which puts crop productivity and food security at risk by increasing the frequency of extreme weather events like heatwaves, floods, and droughts. These unpredictable conditions disrupt growing seasons, reduce soil fertility, and increase pest and disease outbreaks, all of which have an impact on farmers' ability to sustainably maintain output. Additionally, overuse of water and soil resources, especially in regions with intensive agriculture, has hindered sustainable practices by degrading the ecosystem through biodiversity loss, aquifer depletion, and soil erosion.

Another significant impediment to adopting sustainable agricultural practices is the economic and societal barriers that many smallholder farmers face. Due to limited access to markets, capital, technology, and training, many farmers—particularly those in developing countries—often struggle to make the transition to sustainable methods. Short-term financial pressures can occasionally pressure farmers to prioritize monoculture farming or the use of chemical inputs to boost yields over long-term sustainability. Furthermore, a lack of government backing, inadequate policy frameworks, and poor infrastructure may make it difficult for farmers to implement sustainable practices. To overcome these obstacles, more money is needed for financial incentives, education, and infrastructure, particularly in agricultural communities that are already at risk.

Notwithstanding these challenges, certain emerging trends could have an impact on future sustainable food production. Regenerative agriculture, which focuses on restoring soil health, increasing biodiversity, and sequestering carbon, is becoming more popular as a holistic approach to sustainable farming. Alternative proteins such as lab-grown meat and plant-based protein products are becoming more and more popular to reduce the environmental harm caused by animal farming. To feed the rapidly growing urban population, urban agriculture and vertical farming—which produce food closer to consumers while consuming less water and land—are expected to be essential. Furthermore, advancements in agricultural technology like drones, AI-powered precision farming, and genetic editing methods like CRISPR have the potential to boost output, maximize resource use, and improve the resilience of food systems to climate change. If supported by prudent investments and regulations, these advancements have the potential to transform sustainable food production in the future and address the issue of global food security.

9 Conclusion

Sustainable food production is essential for ensuring long-term food security, preserving environmental resources, and addressing the challenges posed by climate change and population growth. By adopting innovative technologies, implementing resource-efficient practices, and fostering community and policy support, sustainable systems can balance the demands of food production with environmental preservation. Collaborative efforts across sectors, including agriculture, industry, and governance, are crucial to achieving these goals. Ultimately, transitioning to sustainable food production not only safeguards the planet's ecosystems but also ensures equitable access to nutritious food for current and future generations.

References

Bale, J., Van Lenteren, J., & Bigler, F. (2008). Biological control and sustainable food production. *Philosophical Transactions of the Royal Society B: Biological Sciences, 363*(1492), 761–776.

Balogun, A.-L., Adebisi, N., Abubakar, I. R., Dano, U. L., & Tella, A. (2022). Digitalization for transformative urbanization, climate change adaptation, and sustainable farming in

Africa: trend, opportunities, and challenges. *Journal of Integrative Environmental Sciences, 19*(1), 17–37.
Boix-Fayos, C., & de Vente, J. (2023). Challenges and potential pathways towards sustainable agriculture within the European green deal. *Agricultural Systems, 207*, 103634.
Brito, L., Bédère, N., Douhard, F., Oliveira, H., Arnal, M., Peñagaricano, F., Schinckel, A., Baes, C. F., & Miglior, F. (2021). Genetic selection of high-yielding dairy cattle toward sustainable farming systems in a rapidly changing world. *Animal, 15*, 100292.
Candemir, A., Duvaleix, S., & Latruffe, L. (2021). Agricultural cooperatives and farm sustainability–a literature review. *Journal of Economic Surveys, 35*(4), 1118–1144.
Chopin, P., Mubaya, C. P., Descheemaeker, K., Öborn, I., & Bergkvist, G. (2021). Avenues for improving farming sustainability assessment with upgraded tools, sustainability framing and indicators. A review. *Agronomy for Sustainable Development, 41*, 1–20.
Dessart, F. J., Barreiro-Hurlé, J., & Van Bavel, R. (2019). Behavioural factors affecting the adoption of sustainable farming practices: A policy-oriented review. *European Review of Agricultural Economics, 46*(3), 417–471.
Dobermann, A., & Nelson, R. (2013). Opportunities and solutions for sustainable food production. In *Sustainable development solutions network.*
Fanzo, J., Haddad, L., Schneider, K. R., Béné, C., Covic, N. M., Guarin, A., Herforth, A. W., Herrero, M., Sumaila, U. R., & Aburto, N. J. (2021). Rigorous monitoring is necessary to guide food system transformation in the countdown to the 2030 global goals. *Food Policy, 104*, 102163.
Gamage, A., Gangahagedara, R., Gamage, J., Jayasinghe, N., Kodikara, N., Suraweera, P., & Merah, O. (2023). Role of organic farming for achieving sustainability in agriculture. *Farming System, 1*(1), 100005.
Gorjian, S., Ebadi, H., Trommsdorff, M., Sharon, H., Demant, M., & Schindele, S. (2021). The advent of modern solar-powered electric agricultural machinery: A solution for sustainable farm operations. *Journal of Cleaner Production, 292*, 126030.
Hamid, M. S., Khalid, D., Noman, I., Ahmed, D. F., Solangi, N. H., Mubarak, N. M., & Jatoi, A. S. (2024). Challenges & perspectives of bio-based additives of food industry. In *Application of bio-additives for the food industry* (pp. 219–239). Springer Nature Switzerland.
Hoek, A. C., Malekpour, S., Raven, R., Court, E., & Byrne, E. (2021). Towards environmentally sustainable food systems: Decision-making factors in sustainable food production and consumption. *Sustainable Production and Consumption, 26*, 610–626.
Jatoi, A. S., & Mubarak, N. M. (2024). *Application of bio-additives for the food industry*. Springer.
Khalid, D., Hamid, M. S., Arsalan, A., Jilani, Y., Hammad, S., Jatoi, A. S., & Mubarak, N. M. (2024). Biochemistry of bio-based additives. In *Application of bio-additives for the food industry* (pp. 29–47). Springer.
Leeuwis, C., Boogaard, B. K., & Atta-Krah, K. (2021). How food systems change (or not): Governance implications for system transformation processes. *Food Security, 13*(4), 761–780.
McKenzie, F. C., & Williams, J. (2015). Sustainable food production: Constraints, challenges and choices by 2050. *Food Security, 7*, 221–233.
Meyer-Aurich, A. (2005). Economic and environmental analysis of sustainable farming practices–a Bavarian case study. *Agricultural Systems, 86*(2), 190–206.
Mpanga, I. K., Schuch, U. K., & Schalau, J. (2021). Adaptation of resilient regenerative agricultural practices by small-scale growers towards sustainable food production in north-Central Arizona. *Current Research in Environmental Sustainability, 3*, 100067.
Muhie, S. H. (2022). Novel approaches and practices to sustainable agriculture. *Journal of Agriculture and Food Research, 10*, 100446.
Ng, K. S., Farooq, D., & Yang, A. (2021). Global biorenewable development strategies for sustainable aviation fuel production. *Renewable and Sustainable Energy Reviews, 150*, 111502.
Nguyen, N., & Drakou, E. G. (2021). Farmers intention to adopt sustainable agriculture hinges on climate awareness: The case of Vietnamese coffee. *Journal of Cleaner Production, 303*, 126828.

Qu, B., Xiao, Z., Upadhyay, A., & Luo, Y. (2024). Perspectives on sustainable food production system: Characteristics and green technologies. *Journal of Agriculture and Food Research, 15*, 100988.

Rust, N. A., Jarvis, R. M., Reed, M. S., & Cooper, J. (2021). Framing of sustainable agricultural practices by the farming press and its effect on adoption. *Agriculture and Human Values, 38*(3), 753–765.

Sattar Jatoi, A., Mubarak, N. M., Solangi, N. H., & Kumar, D. (2024). Introduction to bio-additives. In *Application of bio-additives for the food industry* (pp. 1–13). Springer.

Serebrennikov, D., Thorne, F., Kallas, Z., & McCarthy, S. N. (2020). Factors influencing adoption of sustainable farming practices in Europe: A systemic review of empirical literature. *Sustainability, 12*(22), 9719.

Setsoafia, E. D., Ma, W., & Renwick, A. (2022). Effects of sustainable agricultural practices on farm income and food security in northern Ghana. *Agricultural and Food Economics, 10*(1), 1–15.

Short, R. E., Gelcich, S., Little, D. C., Micheli, F., Allison, E. H., Basurto, X., Belton, B., Brugere, C., Bush, S. R., & Cao, L. (2021). Harnessing the diversity of small-scale actors is key to the future of aquatic food systems. *Nature Food, 2*(9), 733–741.

Slimi, C., Prost, M., Cerf, M., & Prost, L. (2021). Exchanges among farmers' collectives in support of sustainable agriculture: From review to reconceptualization. *Journal of Rural Studies, 83*, 268–278.

Smith, B. G. (2008). Developing sustainable food supply chains. *Philosophical Transactions of the Royal Society B: Biological Sciences, 363*(1492), 849–861.

Smith, P., & Gregory, P. J. (2013). Climate change and sustainable food production. *Proceedings of the Nutrition Society, 72*(1), 21–28.

Trimble, S. (2021). Reducing fresh produce losses and increasing efficiency in harvest & post-harvest. Retrieved March, 26, 2021.

Vittuari, M., Bazzocchi, G., Blasioli, S., Cirone, F., Maggio, A., Orsini, F., Penca, J., Petruzzelli, M., Specht, K., & Amghar, S. (2021). Envisioning the future of European food systems: Approaches and research priorities after COVID-19. *Frontiers in Sustainable Food Systems, 5*, 642787.

Part II
Bioprocessing Techniques

Chapter 4
Application of Bioreactors for Food Processing Industries

Abdul Fatah Soomro, Irfan Ahmed Abbasi, Zenab Tariq Baig, and Abdul Sattar Jatoi

1 Introduction

In the biological process of fermentation, microorganisms such as bacteria, yeast, and fungus use metabolic pathways to convert organic substrates into chemicals that are useful(Bhattacharya et al., 2024). The fact that bread, yogurt, beer, wine, and other foods and beverages have been made using this ancient technique for millennia shows how important it is to human culture and nutrition(Onay & Onay, 2025). However, fermentation serves far more than just traditional functions; it is fundamental to modern biotechnology and serves as the foundation to produce several biochemicals, medications, biofuels, and biopolymers. Fermentation processes, due to their versatility, are essential for meeting the growing demand for sustainable and environmentally friendly manufacturing methods in a rapidly evolving world (Xiaoqi et al., 2025). Fermentation is based on the concept of bioreactors, which are

A. F. Soomro (✉)
Hi-Tech Lab, Department of Enegry and Environmental Engineering, Dawood University of Engineering and Technology, Karachi, Pakistan
e-mail: fatah.soomro@duet.edu.pk; abdul.sattar@duet.edu.pk

I. A. Abbasi
Hi-Tech Lab, Department of Enegry and Environmental Engineering, Dawood University of Engineering and Technology, Karachi, Pakistan

Department of Global Smart City, Sungkyunkwan University, Suwon-Si, Republic of Korea

Z. T. Baig
Department of Environmental Sciences, The University of Haripur, Haripur, Pakistan

A. S. Jatoi
Department of Chemical Engineering, Dawood University of Engineering and Technology, Karachi, Pakistan

A. S. Jatoi (ed.), *Integrated Food and Bioprocessing*,
https://doi.org/10.1007/978-3-031-87209-9_4

specifically designed containers that provide controlled conditions for microbial growth and product generation (Sevda & Kumar, 2024). By optimizing fermentation parameters including temperature, pH, oxygen levels, and nutrient availability, researchers and enterprises can increase the quantity and quality of the desired products (Walsh et al., 2024). The size of bioreactors can vary from tiny laboratory units to massive industrial systems, and they can support a broad range of applications across multiple sectors "With the development and proliferation of bioreactor technology, fermentation processes have become much more efficient and scalable, allowing for innovative solutions in a variety of industries. Presenting a comprehensive examination of fermentation and bioreactor technologies, including their foundations, applications, and prospects, is the aim of this chapter. The many fermentation processes, the biochemical pathways involved, and the construction and operation of bioreactors will all be covered" The chapter will also discuss the challenges that come with fermentation and bioreactor processes, including the necessity to optimize growth conditions, the potential for contamination, and the complexities of downstream processing. Understanding these concepts and trends will help readers understand how important fermentation and bioreactors are to the advancement of biotechnology and the resolution of global concerns like sustainability and resource efficiency (Fig. 4.1).

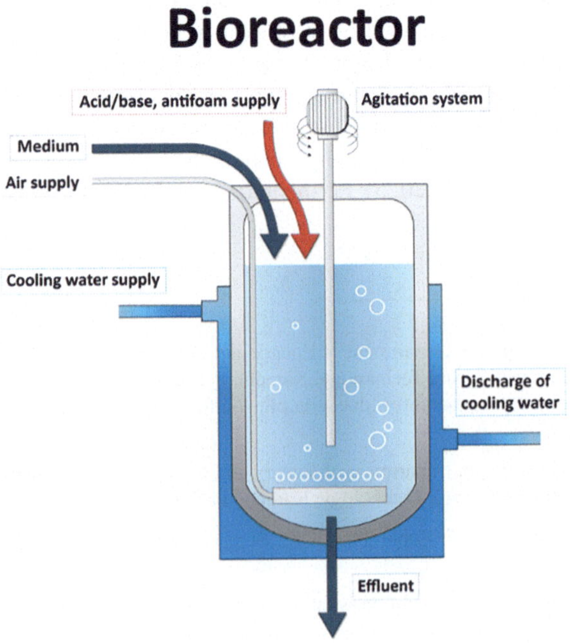

Fig. 4.1 Bioprocessing—fermentation. (Chisti & Moo-Young, 2020)

2 Types of Fermentation

Using bacteria and yeast, fermentation is a metabolic process that converts carbohydrates into acids, gases, or alcohol in an anaerobic setting. It is still necessary in many industrial operations today, having been utilized for thousands of years to manufacture and preserve food. There are other types of fermentation, but the main ones are lactic acid fermentation, acetic acid fermentation, and alcoholic fermentation. A range of food and beverage industries can use each type because to their unique bacteria, metabolic processes, and end products. Alcoholic fermentation is a well-known fermentation process that mostly employs yeasts, particularly *Saccharomyces cerevisiae*. This process turns carbohydrates like fructose and glucose into ethanol and carbon dioxide. Fermentation and glycolysis, which convert glucose to pyruvate, It converts pyruvate into two main steps of the reaction: ethanol and CO_2. This type of fermentation is necessary to generate alcoholic beverages like wine and beer because it converts carbs into alcohol, which enhances the flavor and yields the right amount of alcohol. The resulting carbon dioxide is what gives sparkling wines and beers their carbonation. Alcoholic fermentation is also used in baking, where the CO_2 causes the dough to rise. Another important kind of fermentation that is mostly performed by lactic acid bacteria (LAB) is lactic acid fermentation, which includes with this method, glucose is converted to lactic acid without producing gas. Lactic acid fermentation can be classified into two types: homolactic fermentation, which ferments glucose primarily to produce lactic acid, and heterotactic fermentation, which produces a mixture of lactic acid, ethanol, and carbon dioxide. The production of fermented foods like kimchi and sauerkraut, as well as a range of dairy products including cheese and yogurt, depend on this type of fermentation. Apart from acting as a preservative by inhibiting the development of rotting, the lactic acid produced.

Acetic acid fermentation is the process by which acetic acid bacteria, including *Acetobacter* and *Gluconobacter*, transform ethanol into acetic acid. In contrast to the totally anaerobic fermentations that were previously addressed, this process typically occurs in the presence of oxygen. The first stage of fermentation is called alcoholic fermentation, and it is at this phase that alcohol is produced from carbohydrates" When ethanol is converted to acetic acid by acetic acid bacteria, products such as vinegar are created. To produce various vinegars used in dressing, cooking, and preservation, this type of fermentation is essential in the culinary business. A wide range of culinary applications are made possible by the wide variation in vinegar's flavor, aroma, and acidity, which are contingent upon the ingredients and fermentation circumstances.

In addition to these primary fermentation types, there are several other specialist fermentations. The renowned process of propionic acid fermentation is used to make Swiss cheese, where Propionibacterium species convert lactic acid into propionic acid and carbon dioxide, giving the cheese its distinct flavor and holes. Similarly, butyric acid fermentation, which is often associated with the fermentation of butter and some cheeses, produces butyric acid by Clostridium species. Mixed acid fermentation produces a complex array of end products, including lactic acid, acetic acid, ethanol, and gas. This is observed in several Enterobacteriaceae. Numerous applications are made possible by this versatility, including the production of certain fermented foods and waste treatment techniques.

Lactic acid, acetic acid, alcoholic, and other types of fermentation are defined by distinct microbial communities and metabolic pathways. Each type of fermentation has a specific purpose in the production of food and drink, improving flavor, safety, and preservation. Understanding these different fermentation processes is crucial for optimizing their use in industrial settings, enhancing food quality, and developing innovative products that meet changing consumer demands. As additional research and technological advancements are anticipated to uncover new applications and methods, fermentation's importance in biotechnology and food science will only increase.

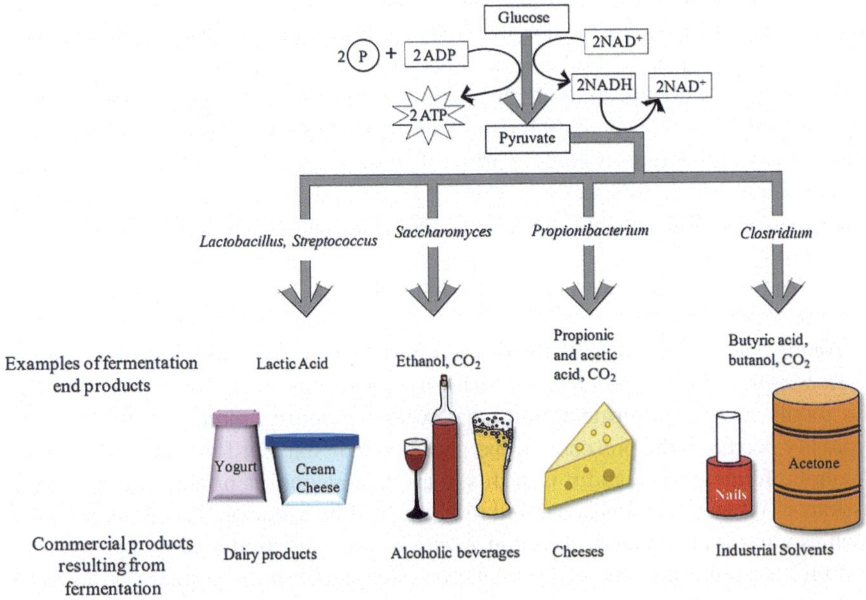

Laboratory exercises in microbiology: discovering the unseen world through hands-on investigation. https://images.app.goo.gl/gAU3A6uogEgmFhu57

3 Biochemical Pathways in Fermentation

Fermentation is a vital metabolic process that allows organisms to generate energy without oxygen. The primary metabolic process in fermentation, glycolysis, converts glucose to pyruvate while simultaneously producing a net increase in ATP (adenosine triphosphate). The energy investment phase and the energy payback phase are the two stages of glycolysis, which takes place in the cytoplasm of cells and is a sequence of enzyme-catalyzed events. Two ATP molecules are used to phosphorylate glucose and transform it into fructose-1,6-bisphosphate during the energy investment phase. The glucose molecule is ready for continued metabolism and eventual cleavage thanks to this mechanism.

Fructose-1,6-bisphosphate degrades into two three-carbon molecules: glyceraldehyde-3-phosphate and dihydroxyacetone phosphate. After this, both molecules can be converted into glyceraldehyde-3-phosphate, which undergoes additional modifications during the energy payout phase.Each glyceraldehyde-3-phosphate molecule is oxidized and phosphorylated to yield ATP and NADH (nicotinamide adenine dinucleotide in its reduced state). Finally, for each glucose molecule, two pyruvate molecules are created, resulting in a net gain of two ATP and two NADH molecules... These compounds are required for the following fermentation activities.

The type of fermentation being used impacts what happens to pyruvate after glycolysis. Pyruvate is decarboxylated to form acetaldehyde and carbon dioxide during the fermentation of alcohol, which is primarily carried out by yeast. Pyruvate decarboxylase is the enzyme that catalyzes this reaction. Following that, acetaldehyde serves as an electron acceptor during the reduction phase, where alcohol dehydrogenase converts NADH back to NAD+. Ethanol, the pathway's final product, can accumulate in the fermentation medium in anaerobic conditions. NAD+ regeneration is critical for maintaining glycolysis and ensuring a consistent supply of ATP.

Pyruvate, on the other hand, is directly converted to lactic acid during the fermentation process, which is predominantly carried out by lactic acid bacteria. Lactate dehydrogenase catalyzes this process, in which NADH gives pyruvate electrons to form lactic acid and replenish NAD+.. Lactic acid fermentation, which occurs in anaerobic settings, is extremely important in the production of foods such as yogurt and fermented vegetables. Unlike alcoholic fermentation, which releases carbon dioxide, this pathway's main output is lactic acid. By lowering the pH of their environment, lactic acid helps preserve fermented foods and gives them a sour taste.

Acetic acid fermentation is a unique process that begins when ethanol is converted to acetic acid. Acetic acid bacteria use ethanol as a substrate and produce acetic acid through a two-step conversion process. Ethanol is first oxidized to acetaldehyde, which is subsequently converted to acetic acid by enzymes like alcohol dehydrogenase and aldehyde dehydrogenase, respectively. This process is oxygen-dependent and occurs in the presence of oxygen, which sets it apart from the fully anaerobic pathways of lactic acid fermentation and alcoholic fermentation. Acetic acid generation not only gives vinegar its characteristically sour flavor but also acts as a preservative by inhibiting the growth of bacteria.

Different outputs from different fermentation methods reveal different metabolic pathways at play. For instance, Clostridium species produce butyric acid, hydrogen, and carbon dioxide by the fermentation of lactic acid or carbohydrates. This process gives various cheeses their distinctive qualities and is essential to the creation of numerous dairy products. Lactic acid, acetic acid, ethanol, formic acid, and hydrogen gas are among the complex mixture of acids, alcohols, and gases produced by mixed acid fermentation, which is present in enteric bacteria and demonstrates the variety of metabolic pathways used by various microorganisms.

Fermentation routes need to be controlled to increase productivity and efficiency. Crucial regulatory enzymes, like phosphofructokinase in glycolysis, react to the energy level of the cell, enabling the organism to modify its metabolic pathway in response to available substrates and environmental changes. For instance, in environments with high glucose levels, fermentation routes may be preferred, but in environments with high oxygen levels, aerobic respiration may take over because it generates far more ATP than fermentation alone. The flexibility of their metabolic pathways allows organisms to thrive in a range of environments with varying substrate availability.

Variables including temperature, pH, and nutrient availability can have a significant impact on the growth of microorganisms and the efficiency of fermentation processes. These factors also have an impact on fermentation pathways. Temperature variations can alter the activity of enzymes, but pH affects microbial metabolism and the stability of fermentation products. Additionally, the rate at which fermentative bacteria grow and, consequently, the yield of the finished product may be affected by the availability of nutrients, particularly supplies of nitrogen and phosphorus. Therefore, understanding these environmental factors is crucial for optimizing industrial fermentation operations.

In summary, the biochemical pathways involved in fermentation are a web of interconnected metabolic processes that convert carbohydrates into a range of end products, including alcohol, gasses, and acids. Each type of fermentation has distinct pathways that show the range of metabolic processes present in nature and the abilities of the microbes involved. Biotechnology and fermentation technology advancements continue to explore these pathways, increasing yields, creating new products, and expanding our knowledge of microbial metabolismWith the advancement of fermentation research, sustainable practices in food production, biofuel production, and bioremediation initiatives could potentially address contemporary environmental and health concerns.

4 Bioreactors: Overview and Types

Producing biological products such as enzymes, antibodies, vaccines, and biofuels is the primary function of bioreactors, which are controlled environments where biological activities occur. They produce an environment that is favorable to the growth and health of microorganisms or cells by optimizing factors like temperature, pH, oxygen levels, and nutrient availability. Because their design and operation are crucial to ensuring the efficacy and output of the intended products, bioreactors are vital components of biotechnological processes in a range of industries, from food production to pharmaceuticals.

Batch bioreactors are one of the simplest types of bioreactors; they work in separate batches rather than continuously. All of the ingredients, such as the nutrients and inoculum, are added at the beginning of the process in a batch bioreactor. The products are gathered and the reactor is cleaned and prepared for the next batch after a predetermined period of time for the reaction to finish. This type of bioreactor is widely used in processes that need high levels of product quality and purity, such as the manufacturing of antibiotics and other medications. Despite its greater simplicity of use and control, batch bioreactors can have a lower output capacity and require more downtime for cleaning and sterilizing.

Alternatively, continuous bioreactors operate continuously, allowing for a constant flow of nutrients into the system and removing waste at the same time. A more stable environment for microbial development is provided by this technology since circumstances can be sustained throughout time. Continuous bioreactors are highly advantageous for large-scale production processes where efficiency and throughput are critical, such as the manufacture of fermentation products or biofuels. These bioreactors require close monitoring and management in order to prevent contamination and preserve optimal conditions for the microorganisms involved.

A continuous bioreactor subtype that combines the features of batch and continuous systems is the fed-batch bioreactor. This arrangement is used to supply the initial batch of nutrients, and additional nutrients are introduced to the reactor as the fermentation process goes on. This approach gives you greater control over the concentration of nutrients, which optimizes growth conditions and product development. Fed-batch systems are commonly used to create high-value products such as medicines and enzymes, where precise control over the growth environment is necessary to achieve high yields and quality.

Stirred-tank bioreactors are among the most widely used models for both batch and continuous processing. The cylindrical tank of these reactors is equipped with an agitator to promote mixing and oxygen transmission, both of which are essential for aerobic fermentation processes. The agitation improves heat and mass transfer efficiency by guaranteeing that nutrients and microorganisms are dispersed evenly throughout the medium. Stirred-tank bioreactors can be operated at a range of scales, from laboratory to industrial, and can be configured with advanced monitoring and control systems to optimize fermentation conditions.

The **airlift bioreactor** is another type of bioreactor that employs a different mixing principle. In airlift reactors, a gas is introduced into the liquid medium to produce a flow pattern that circulates the liquid without the need for mechanical agitation. This method creates a softer environment for fragile cells or microorganisms while also improving oxygen transfer. For the growth of shear-sensitive cells, such as the mammalian cells required to produce vaccines, airlift bioreactors are particularly useful. Their shape facilitates efficient gas-liquid interaction and lessens the chance of agitation-induced cell damage.

Membrane bioreactors (MBRs) are integrated systems that combine biological treatment procedures with membrane filtration methods for the treatment of wastewater and the recovery of bioproducts. MBRs maintain microorganisms in the reactor while enabling the selective removal of products through a membrane. This arrangement increases microbial concentrations and improves treatment efficacy by enhancing biomass retention. MBRs are increasingly being used to treat municipal and industrial wastewater as well as to recover valuable by-products from fermentation activities.

The type of bioreactor that is selected depends on the specific requirements of the biological process, the characteristics of the microorganisms or cells being used, and the desired outcome. While batch or fed-batch systems might be better suited for operations that demand tight sterility, stirred-tank or air-field bioreactors might be more beneficial for those that require high oxygen transfer rates. Furthermore, the ability to scale from laboratory-scale to industrial-scale production can have a significant impact on the success of a biotechnological venture, making the bioreactor's scalability a crucial consideration.

One of the ongoing challenges in the design and operation of bioreactors is maintaining optimal conditions for microbial growth while preventing contamination. For bioreactor systems to ensure that the desired microorganisms can thrive free from unwanted contaminants, rigorous sterilization and cleaning protocols are necessaryA stable environment and optimal product yields depend on the constant monitoring of critical factors such as temperature, pH, dissolved oxygen, and nutrient concentrations. In order to facilitate real-time condition monitoring and change, bioreactor systems are gradually integrating automation and control technology improvements.

The efficient synthesis of a wide range of biological products is made possible by bioreactors, which are crucial components of biotechnology. Batch, continuous, fed-batch, stirred-tank, airlift, and membrane bioreactors are among the various types of bioreactors; each has special advantages and applications that are appropriate for specific processes. As the field of biotechnology develops, improvements in bioreactor design and operation are expected to lead to increased efficiency, sustainability, and scalability in order to meet the growing demand for bioproducts in sectors ranging from healthcare to food and environmental management. Understanding the intricacies of bioreactor systems will be essential to making the most of them and advancing biotechnological applications in the future (Fig. 4.2).

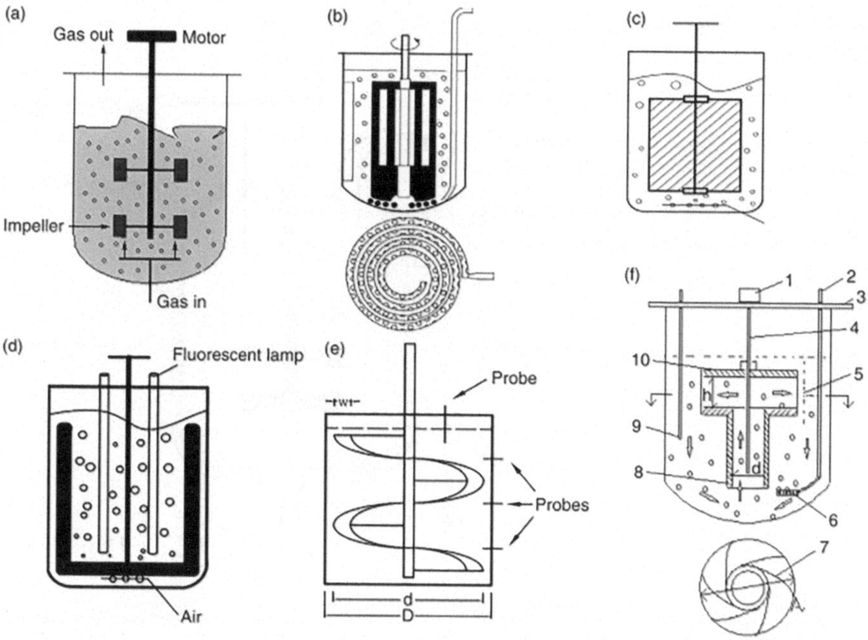

Fig. 4.2 Encyclopedia of physical science and technology (Third Edition). (Suslick, 2001)

5 Design and Operation of Bioreactors

Important aspects of biotechnology that impact the effectiveness and productivity of biological processes are the construction and functioning of bioreactors. A well-designed bioreactor provides the conditions necessary for microorganisms or cells to grow and produce the desired products, and effective operation ensures that these parameters are maintained throughout the fermentation process. To optimize bioprocess performance, critical design elements such as reactor configuration, material selection, temperature control, pH regulation, oxygen transfer, and mixing methods must be appropriately considered (Fig. 4.3).

5.1 *Reactor Configuration and Materials*

The growth and production of the required metabolites by microorganisms can be significantly impacted by the configuration of a bioreactor. Common forms include stirred-tank reactors, airlift reactors, and membrane reactors; depending on the application, each offers special advantages. Selecting the right material is also very important; most bioreactors are constructed of glass, stainless steel, or specialty

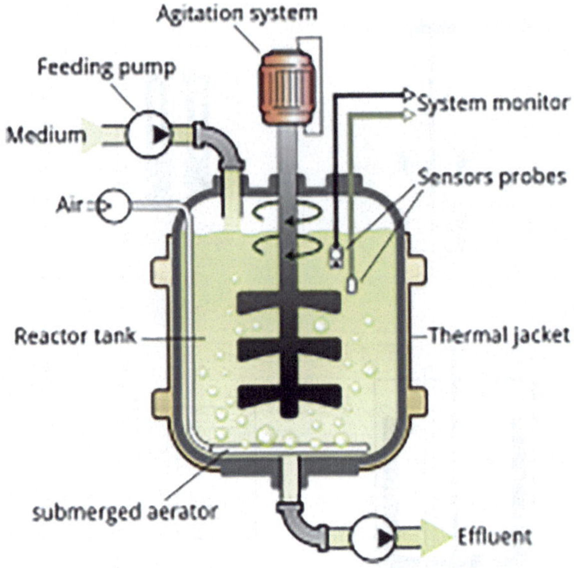

Fig. 4.3 Bioreactor. (Suslick, 2001)

polymers that are resistant to corrosion at high pressures and temperatures. These components must also be compatible with the cellular or microbiological systems being employed to prevent contamination and ensure product integrity.

5.2 Temperature Control

Temperature is a critical component in bioreactor operation because it affects microbial metabolism, growth rates, and product output. Most bacteria have optimal temperature ranges that must be maintained for fermentation to occur. The temperature in bioreactors is managed by heat exchangers, internal coils, or heating or cooling jackets. The optimal temperature is maintained with the help of automated control systems and continuous monitoring, allowing for prompt modifications in reaction to changing conditions. Poor temperature control could lead to reduced yields or the production of unwanted byproducts.

5.3 pH Regulation

pH levels have a significant effect on both the activity of enzymes and the growth of microbes. Most biological systems operate best in pH ranges, and deviations can negatively affect metabolism and the production of new products. Bioreactors typically contain pH sensors and automated titration systems that add bases or acids to maintain the right pH. The use of buffering agent in the growth medium can also stabilize pH values during fermentation. Regular pH monitoring and necessary adjustments are required to optimize yields and ensure consistent product quality.

5.4 Oxygen Transfer

In aerobic fermentation systems, oxygen transport is a key factor affecting microbial growth and metabolism. The rate of oxygen transfer depends on several factors, such as the design of the bioreactor, the speed of agitation, and the amount of dissolved oxygen in the mediumWhile stirred-tank bioreactors usually use agitation to boost oxygen transfer rates, airlift reactors use gas injection to promote mixing and aeration. By measuring dissolved oxygen levels, sensors allow for real-time adjustments to agitation or aeration rates, ensuring that microorganisms receive adequate oxygen to thrive.

5.5 Mixing Strategies

Bioreactor operation depends on effective mixing to ensure that nutrients, cells, and gases are dispersed evenly throughout the growth medium. Inadequate mixing may lead to localized nutrient depletion, waste product accumulation, and uneven growth rates among microorganisms. Various mixing methods, such as mechanical agitation, pneumatic mixing, and static mixers, may be employed, depending on the kind and size of the reactor. Computational fluid dynamics (CFD) simulations are widely used in the design phase to improve reactor mixing patterns and mass transfer rates.

5.6 Control Systems

Advanced control systems are necessary to monitor and regulate critical parameters due to the complexity of bioreactor operation. Automated control systems that integrate a range of sensors and actuators continuously monitor temperature, pH, dissolved oxygen, and nutrient concentrations. Real-time feedback can be obtained by analyzing the data from these sensors, allowing for automated or human adjustments. Two examples of advanced control methods that are increasingly being employed to increase the precision and efficiency of bioprocess operations are model predictive control (MPC) and feedback loops.

5.7 Scale-Up Considerations

There are particular challenges when expanding bioreactors from laboratory to industrial scale. Variables including mass movement, heat transmission, and mixing dynamics can alter significantly at larger scales, necessitating adjustments to design and operation. Pilot studies may be conducted as part of scale-up plans to evaluate performance prior to full-scale production. Engineers must consider how changes to reactor dimensions, geometry, and operating conditions may impact biological performance to prevent scalability from compromising product quality or yield.

5.8 Quality Assurance and Sterilization

Sterility must be maintained during bioreactor operations to prevent contamination by unwanted microbes, which can reduce product quality and yield. Chemical sterilization techniques like steam sterilization (autoclaving), hydrogen peroxide, or ethanol are frequently used in bioreactors, which frequently have built-in sterilizing capabilities. To make sure that every part of the bioreactor is thoroughly sterilized prior to inoculation, the sterilization procedure needs to be verified. Aseptic techniques are also required during the fermentation and inoculation processes to lower the risk of infection.

6 Applications of Bioreactor

In many commercial and research applications, fermentation and bioreactors are essential because they leverage the metabolic powers of cells, yeast, and microbes to create a variety of products. Bioreactor technology and fermentation processes have numerous and significant applications in the production of food, drinks, medicines, and biofuels.. Given these applications, it becomes evident how crucial fermentation is to modern biotechnology and how it may inspire future advancements (Fig. 4.4).

6.1 Food and Beverage Production

Fermentation is most commonly used in the food and beverage industry. Among the staples that fermentation produces are bread, cheese, yogurt, and fermented beverages like wine and beer. These processes often convert sugars into acids, alcohols, or gasses that enhance flavor, texture, and preservation using lactic acid bacteria, yeast, and molds. For instance, during the yogurt-making process, Lactobacillus

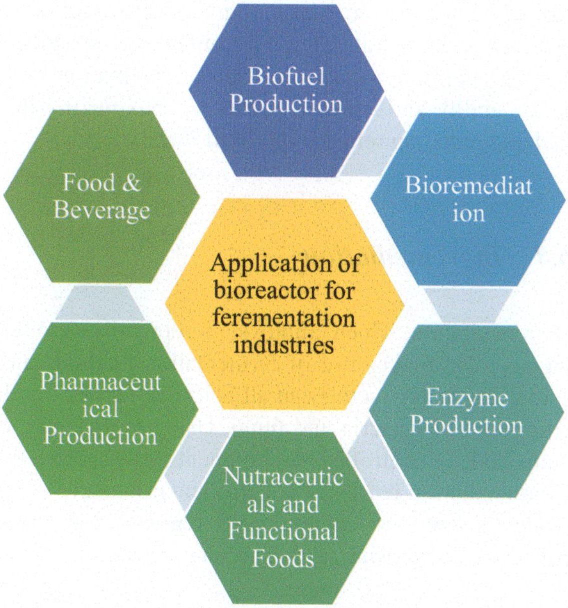

Fig. 4.4 Application of bioreactor for various fermented products formation

species digest lactose into lactic acid, lowering pH and adding a sour taste while maintaining the finished product. In a similar vein, yeast ferments the carbohydrates in grains or fruit using a centuries-old method to produce alcohol in beverages.

6.2 Pharmaceutical Production

Fermentation processes are crucial to the pharmaceutical industry's production of vaccines, antibiotics, and other bioactive compounds. Microorganisms like *Penicillium and Streptomyces* are used in bioreactors to ferment antibiotics like penicillin and streptomycin. Additionally, bioreactors facilitate the production of viral antigens from mammalian cells as well as the cultivation of cells for vaccine manufacturing. The ability to scale up fermentation processes in bioreactors allows for the controlled and efficient production of these essential therapeutic chemicals, meeting the requirements of the pharmaceutical market.

6.3 Biofuel Production

The production of biofuels like ethanol and biodiesel is a significant use of fermentation and bioreactors. Yeast and bacteria are utilized in fermentation processes to convert sustainable biomass, such corn, sugarcane, or agricultural waste, into biofuels. For example, Saccharomyces cerevisiae is commonly used in the fermentation of carbohydrates derived from corn to produce ethanol, a well-liked biofuel. Through continuous and controlled fermentation, the use of bioreactors enhances sustainability and energy security by maximizing the conversion of biomass into fuel and reducing reliance on fossil fuels.

6.4 Bioremediation

Bioremediation, which employs microorganisms to break down environmental pollutants, also makes use of fermentation methods and bioreactors. Pesticides, oil spills, and heavy metals can all be broken down into less dangerous forms by specific types of bacteria and fungi. Using bioreactors, these microorganisms can be cultivated in controlled environments to increase their capacity for breakdown. Bioremediation reduces the detrimental impacts of human activity on ecosystems while cleaning up contaminated regions in a sustainable and environmentally friendly way by using fermentation processes.

6.5 Enzyme Production

In several industries, including food processing, detergents, and biofuels, fermentation is an essential method for producing enzymes. Certain enzymes, such as cellulases, amylases, and proteases, are produced by microorganisms like bacteria and fungus through fermentation processes in bioreactors. These enzymes' capacity to convert complex substrates into simpler molecules boosts industrial activities' efficiency. For example, the food industry uses cellulases to manufacture bioethanol from lignocellulosic biomass and fermentation-produced amylase to convert starch.

Enzymes produced through fermentation are used extensively in a range of industrial operations, such as paper production, textile processing, bio-remediation, as well as in food and medicine. In manufacturing operations, fermentation-produced enzymes can replace harsh chemicals, making them more environmentally friendly. For instance, to improve the efficiency of the pulping process, the paper industry treats pulp with xylanases and cellulases made from fungal fermentation. The versatility of fermentation-produced enzymes opens up new avenues for eco-friendly business practices across a range of industries.

6.6 Nutraceuticals and Functional Foods

The production of functional meals and nutraceuticals—which provide health benefits beyond mere sustenance—is increasingly utilizing fermentation. Probiotics are live bacteria that, when ingested, have health advantages and are made using fermentation procedures. High viability and efficacy in the final product are ensured by the controlled fermentation of probiotic strains in bioreactors. Additionally, fermentation can improve a nutrient's bioavailability, which facilitates the body's absorption of it. This application illustrates how fermentation might enhance diet-related health and well-being.

Fertilizer and animal feed are produced in agriculture through fermentation. Fermented diets can improve nutritional availability and digestibility for animals. Additionally, fermentation techniques can yield microbial inoculants that promote plant development and enhance soil health. These inoculants can promote sustainable farming practices by improving disease resistance and nutrient uptake. Bioreactors are used to scale up the production of these beneficial bacteria, ensuring their effectiveness in increasing agricultural produce.

6.7 Biomaterials Production

Biopolymer and biomaterial synthesis based on fermentation is gaining popularity for a variety of applications, including biodegradable polymers and scaffolds for tissue engineering. Microorganisms can be altered to produce biopolymers such as polylactic acid (PLA) or polyhydroxyalkanoates (PHA) through fermentation processes. The sustainable manufacturing of biodegradable materials that can take the place of traditional plastics is made possible by the large-scale growth of these bacteria made possible by bioreactors. The potential advantages of fermentation in resolving environmental issues related to plastic waste are demonstrated by this application.

Finally, because they allow researchers to study microbial metabolism, enzyme function, and genetic engineering, bioreactors and fermentation processes are essential for research and development. By modifying variables like temperature, pH, and nutrient supply, researchers can more easily study metabolic pathways and product generation under controlled fermentation conditions. The development of novel bioprocesses and products using bioreactors enables advances in biotechnology. The significance of fermentation and bioreactors for the development of science and technology is highlighted by this application.

7 Challenges in Fermentation and Bioreactor Processes

Fermentation and bioreactor technologies, which offer countless opportunities for producing lucrative goods in a range of industries, have revolutionized biotechnology. However, these processes also face a number of challenges that could impact production, productivity, and product quality. Understanding these challenges is necessary to optimize bioprocesses and ensure their efficacy in industrial applications. Below is a list of some of the primary problems with fermentation and bioreactor operations. One of the main challenges in fermentation and bioreactor operations is the potential for contamination by unwanted bacteria. Contamination may lead to reduced yields, altered product quality, and increased manufacturing costs since more purification processes are needed. If impurities outcompete the targeted microorganisms for nutrients, the fermentation process may fail. Maintaining sterile conditions throughout inoculation and the fermentation process is crucial, even though achieving and maintaining sterility in large-scale bioreactors can be challenging. Strict adherence to aseptic protocols and effective sterilizing techniques are necessary to lessen these risks. Establishing and maintaining optimal growing conditions for microorganisms or cells is necessary to maximize productivity and yield in fermentation operations. Temperature, pH, nutrient concentrations, and dissolved oxygen levels are among the variables that must be carefully controlled and monitored. However, variations in these parameters could result from equipment failures or bioprocess modifications, which could negatively affect microbial metabolism and product production. Smaller businesses may not always have access to or be able to afford the advanced monitoring and control technologies that these systems require due to their complexity. Extending fermentation processes from the laboratory or pilot size to the industrial scale presents unique challenges. Variables that impact microbial growth and product generation, like mass transfer, heat transfer, and mixing dynamics, may show different behaviors at larger scales. Uneven temperature distribution or insufficient oxygen transport brought on by a larger reactor volume, for example, could result in lower product yields. Comprehensive pilot studies and reactor design modifications are sometimes required to solve these scale-up challenges, which can be costly and time-consuming. Numerous fermentation procedures may lead to the buildup of metabolic byproducts that prevent microbes from growing and functioning, hence reducing process efficiency and product yield. For example, high ethanol concentrations during ethanol production may harm yeast cells, lowering productivity and fermentation rates. By-product concentrations must be controlled by process optimization, such as fed batch or continuous fermentation methods, in order to lessen inhibitory effects. However, this makes the process more difficult and may require additional instruments or adjustments to the way things are done. The economic viability of fermentation and bioreactor processes is greatly influenced by production costs and resource availability. The cost of raw materials, such as substrates and fertilizers, can significantly impact on the overall cost of manufacturing. Large-scale operations may also require a lot of energy for heating, cooling, and aeration. A key challenge for bioprocess engineers is to balance these costs while maintaining high productivity and product quality. By using strategies to optimize resource use, including recycling nutrients or utilizing waste substrates, these budgetary constraints can be mitigated. The microorganisms used in fermentation processes must be genetically stable in order to produce and produce products with consistent quality. However, microorganisms may undergo genetic alterations over time that lead to variations in metabolic activity and product production. These changes can be caused by mutations, plasmid loss, and environmental stress. Monitoring the genetic stability of microbial strains and employing strategies to maintain or enhance stability, such as the use of controlled conditions and rigorous selection procedures, are necessary for dependable fermentation results. Following the completion of fermentation, downstream processing is required to separate and purify the desired products from the fermentation broth. Among the numerous procedures that can make this stage difficult, and resource-intensive are filtration, centrifugation, chromatography, and crystallization. Efficiency and cost of these processes can have a significant effect on the fermentation company's overall financial viability. Additionally, the presence of impurities or byproducts may complicate purification, leading to reduced yields and increased costs. By streamlining downstream processing, integrated bioprocessing technology can address these problems. Temperature fluctuations, humidity, and air quality are examples of environmental variables that might impact fermentation processes and bioreactor performance. For example, variations in the ambient temperature can impact the temperature within bioreactors, changing the way microorganisms metabolize and produce their products. Additionally, contaminants in the air or water source may allow unwanted chemicals or bacteria to infiltrate the fermentation system. Strict environmental controls and monitoring procedures must be implemented in order to mitigate these impacts and ensure consistent bioprocess performance. Even though process efficiency has grown due to advancements in fermentation technology and bioreactor design, there are still certain limitations. For example, standard batch fermentation may not be suitable for all applications due to its limitations in terms of productivity and scalability. The intricacy of bioreactor systems may also mean that smaller businesses lack the specialized knowledge and training required for operators. It will take constant innovation and investment in new technology, such advanced sensors, automation, and data analytics, to overcome these limitations and enhance fermentation processes. To ensure the effectiveness, safety, and quality of the finished products, the fermentation and bioprocessing industries are subject to stringent regulations. Following these rules could be challenging, particularly for new products or processes. Regulations may call for extensive testing, documentation, and validation of fermentation processes—tasks that can be costly and time-consuming. Navigating diverse regulatory environments requires careful planning, in-depth understanding of the relevant rules, and collaboration with regulatory agencies in order to ensure compliance and successful product clearance.

8 Future Trends in Fermentation and Bioreactor Technologies

The subject of fermentation and bioreactor technologies is rapidly evolving due to advancements in research, engineering, and sustainability. Future fermentation and bioprocessing methods may be influenced by a variety of new themes as industries realize the potential of biotechnology to address global problems. These trends include the development of new products, sustainable practices, and technical developments that aim to reduce their adverse environmental effects and boost production. Precision fermentation, a technology that uses state-of-the-art biotechnological techniques to precisely manufacture certain components and compounds, is gaining popularity. GMO microorganisms are commonly used in this process to produce complex molecules like as proteins, flavors, and bioactive compounds.. Utilizing metabolic engineering and synthetic biology, precision fermentation modifies microbial pathways to optimize yield and reduce waste. This approach has the potential to create sustainable alternatives to animal-based products, like dairy proteins and meat substitutes, in response to the growing demand for plant-based and cell-based diets. The integration of automation and digital technologies into the design and operation of bioreactors is expected to improve process efficiency and dependability. The increasing use of advanced sensors, data analytics, and machine learning algorithms to measure fermentation parameters in real time enables predictive analytics and adaptive control tactics. This shift to intelligent bioprocessing not only enhances growth conditions and product yields but also reduces the need for manual intervention, allowing for greater scalability and cheaper operational costs. Digital twins, or virtual representations of bioreactor systems, are also being studied to simulate and optimize bioprocesses before they are implemented. The fermentation and bioprocessing industries are moving toward greener practices that put an emphasis on resource efficiency and waste reduction. This trend includes the use of waste substrates, such as food and agricultural wastes, as feedstocks for fermentation processes. The circular economy can be supported by employing bioconversion techniques to transform these waste materials into usable products like animal feed, biofuels, and bioplastics. Two examples of environmentally friendly bioprocessing methods that are gaining popularity include non-toxic solvents and green chemistry approaches, which lessen the ecological impact of fermentation processes. Fermentation advances in the future will continue to yield complex microbial strains with enhanced performance characteristics. CRISPR-Cas9 and other gene-editing technologies allow for precise modification of microbial genomes, resulting in strains that are more resilient to environmental stressors, develop more quickly, and produce more. Additionally, techniques like as adaptable laboratory evolution (ALE) will be used to create robust strains that can thrive in industrial fermentation environments. This enhanced strain development will be crucial for bioprocess optimization and consistent product quality. The study of multi-organism fermentation systems is a new idea that capitalizes on the cooperative relationships between multiple microbial species. By combining complementary metabolic pathways, multi-species fermentation can improve the production of desired metabolites and decrease the accumulation of hazardous byproducts. By mimicking natural ecosystems, this technique can lead to more effective synthesis of complicated compounds, such as medications and biofuels. It is projected that more microorganisms will be co-cultured in bioreactors, opening up new avenues for bioprocessing innovation. Combining fermentation and bioreactor technologies with other biotechnological approaches, such as cell culture systems and enzyme technology, will drive advancements in bioprocessing. When combined, fermentation and enzymatic processes can enhance product purification and recovery while streamlining downstream processing. Additionally, the development of hybrid bioprocessing systems that integrate fermentation and cell culture techniques would enable the production of complex biomolecules, such as therapeutic proteins and antibodies, with increased yields and purity.

The efficiency and economics of product recovery and purification are expected to be enhanced by advancements in downstream processing technologies. Membrane filtering, chromatography, and extraction techniques are some of the innovative approaches being developed to expedite purification processes and minimize product loss. Innovations in continuous processing and integrated bioprocessing techniques will reduce the time and resources required for downstream processing, improving the overall efficiency of bioreactor and fermentation operations. Consumer desire for functional and health-promoting foods is driving research into fermentation technologies that enhance the nutritional profile of products. Probiotics, prebiotics, and other fermentation-produced bioactive compounds are gaining popularity due to their potential health benefits. Future trends are likely to focus more on developing fermentation techniques that produce useful components in order to meet consumer expectations for health-conscious products. This includes more traditional fermented meals as well as novel formulations that include probiotics and bioactive ingredients in everyday products. Bioreactor designs are expected to evolve throughout time to include new materials and technologies that improve performance and sustainability. Innovations in reactor design, such as modular bioreactors and microfluidic devices, will make bioprocessing more adaptable and scalable. Additionally, the proliferation of bioreactors built from recyclable or biodegradable materials might help achieve sustainability goals. Advanced mixing methods like ultrasonic or vertexing can further improve mass transfer and reduce energy usage during fermentation. Innovation in the quickly growing sectors of fermentation and bioprocessing will require international collaborations between academic institutions, research groups, and businesses.. Cooperative research projects will help improve fermentation processes by addressing common problems, sharing best practices, and developing new technologies. This trend will ensure that global stakeholders may employ fermentation technology to meet the need for sustainable products both domestically and internationally, while also encouraging information sharing and capacity building.

9 Conclusion

In conclusion, fermentation and bioreactor technology have a bright future because to precision fermentation, automation, sustainability, and the development of new products. The increasing application of biotechnology by industry to address consumer demands and environmental concerns will influence the next generation of fermentation processes. As a result, the bioprocessing environment will become more sustainable and efficient. In order to fully utilize fermentation technologies in the years to come, ongoing research and development efforts as well as a focus on collaboration will be necessary.

References

Bhattacharya, R., Arora, S., & Ghosh, S. (2024). Bioprocess optimization for food-grade cellulolytic enzyme production from sorghum waste in a novel solid-state fermentation bioreactor for enhanced apple juice clarification. *Journal of Environmental Management, 358*, 120781.

Chisti, Y., & Moo-Young, M. (2020). Fermentation technology, bioprocessing, scale-up and manufacture. In *Biotechnology-the science and the business* (pp. 177–222). CRC Press.

Onay, M., & Onay, A. (2025). Chapter 41 – Bioethanol production in algal bioreactors. In Y. Gagnon, E. Jacob-Lopes, L. Q. Zepka, & M. C. Deprá (Eds.), *Algal bioreactors* (pp. 639–648). Elsevier Science Ltd.

Sevda, S., & Kumar, S. (2024). Chapter 1 – Introduction of bioreactor. In S. Sevda & S. Kumar (Eds.), *Bioreactor design concepts for viral vaccine production* (pp. 1–34). Academic.

Suslick, K. S. (2001). *Encyclopedia of physical science and technology. Sonoluminescence and sonochemistry* (3rd ed., pp. 1–20). Elsevier Science Ltd.

Walsh, É., Margassery, L. M., Rodriguez-Sanchez, A., Wall, D., Bolger, P., Jansen, M. A. K., & O'Leary, N. (2024). Integration of microbial bioreactors and Lemna minor cultivation for sustainable treatment of dairy processing wastewater. *Journal of Water Process Engineering, 67*, 106290.

Xiaoqi, C., Kang, N., & Pei, X. (2025). Enhancing medicinal proteins production in plant bioreactors: A focal review on promoters. *Fitoterapia, 180*, 106338.

Chapter 5
Enzymatic Conversion

Abdul Sattar Jatoi and Arshad Iqbal

1 Introduction

In biological reactions, enzymatic conversion refers to the process by which enzymes function as catalysts to assist transform substrates into products. Numerous biological processes, including metabolism, digestion, and biosynthesis, rely on this phenomenon. Enzymes—typically proteins—can catalyze reactions in mild conditions with neutral pH and ambient temperatures due to their remarkable efficiency and selectivity. Due to this unique ability, enzymatic conversion is not only essential for life but also plays a significant role in a variety of industrial processes, including the manufacturing of food, medications, and biofuels. The significance of enzymatic conversion in biotechnology cannot be overstated. It uses less energy and generates fewer byproducts than traditional chemical processes, enabling the exact synthesis of high-value compounds. As industries strive to implement more sustainable and environmentally friendly practices, enzymatic conversion has emerged as a crucial strategy. Enzymes can reduce the environmental effect of production processes and promote the concepts of the circular economy by turning waste resources into valuable items. In order to provide a comprehensive grasp of enzymatic conversion, this chapter will look at its mechanics, several types of reactions, and the various sources of enzymes. It will examine the kinetics of enzymes, production processes, and purification techniques in addition to discussing the advancements in enzyme technology that expand its potential applications. The several industries that use enzymatic conversion, the challenges faced in the sector, and the future advances that could further progress this crucial biotechnology topic will also be covered in this chapter. As a result of this inquiry, readers will have a better

A. S. Jatoi (✉) · A. Iqbal
Department of Chemical Engineering, Dawood University of Engineering and Technology, Karachi, Pakistan
e-mail: abdul.sattar@duet.edu.pk

A. S. Jatoi (ed.), *Integrated Food and Bioprocessing*,
https://doi.org/10.1007/978-3-031-87209-9_5

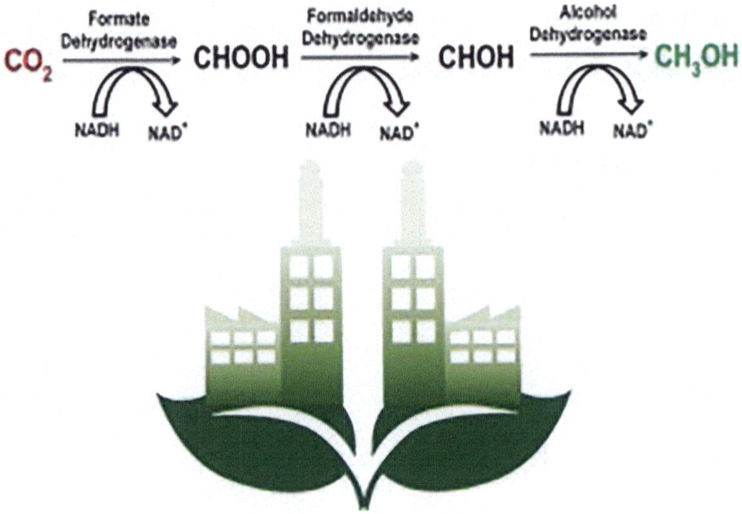

Fig. 5.1 Enzymatic conversion of CO_2 to CH_3OH via reverse dehydrogenase cascade biocatalysis. (Marpani et al., 2017)

understanding of the vital role that enzymatic conversion plays in shaping modern biotechnological processes and addressing global concerns (Fig. 5.1).

2 Enzyme Production and Characterization

Enzymes play a pivotal role in food and bioprocessing industries, acting as biocatalysts that drive essential biochemical reactions with high specificity and efficiency. Understanding enzyme production and characterization is fundamental for optimizing their application in industrial processes. This discussion explores the sources, production techniques, isolation methods, and characterization processes critical to enzyme utilization.

2.1 Sources of Enzymes

Enzymes are derived from three primary sources: microorganisms, plants, and animals. Among these, microbial sources are the most commonly utilized due to their versatility, scalability, and cost-effectiveness. Bacteria, fungi, and yeast are extensively employed for enzyme production because they can be genetically modified to enhance yield and functionality. For instance, "Aspergillus niger" and "Bacillus subtilis" are widely used for producing amylases and proteases.

Plant-derived enzymes, such as papain from papaya and bromelain from pineapple, are valued for their specificity in applications like tenderizing meat and clarifying beverages. Animal-derived enzymes, including rennet for cheese production and trypsin for protein hydrolysis, play critical roles in food and pharmaceutical industries. However, concerns about ethical sourcing and stability have led to increased reliance on microbial enzymes.

2.2 Techniques for Enzyme Production

The production of enzymes involves either submerged fermentation (SmF) or solid-state fermentation (SSF). Submerged fermentation, where microorganisms grow in a liquid nutrient medium, is preferred for industrial-scale production due to ease of process control and automation. SmF is particularly suitable for enzymes like lipases and cellulases. On the other hand, SSF involves microbial growth on solid substrates, making it ideal for producing enzymes like pectinases and tannases, particularly in regions with agricultural by-products as substrates.

Optimization of fermentation parameters such as temperature, pH, aeration, and nutrient composition is essential to maximize enzyme yields. Recent advances in bioprocessing technologies, such as the use of genetically engineered strains and bioreactors with real-time monitoring, have significantly improved enzyme productivity. Additionally, the integration of omics technologies (genomics, proteomics, and metabolomics) enables a deeper understanding of microbial systems, facilitating the design of highly efficient enzyme production processes.

2.3 Enzyme Isolation and Purification

The downstream processing of enzymes involves several steps: cell disruption, centrifugation, filtration, and chromatography. Cell disruption is crucial for intracellular enzymes and can be achieved mechanically (e.g., sonication, bead milling) or chemically (e.g., using detergents). Extracellular enzymes, secreted into the fermentation medium, are more accessible and require less intensive recovery procedures(Liang et al. 2024).

Centrifugation and filtration separate cell debris from the enzyme-rich supernatant. Further purification is achieved through techniques like precipitation (e.g., ammonium sulfate precipitation), ultrafiltration, and chromatographic methods such as ion-exchange, affinity, and size-exclusion chromatography. Purity levels required depend on the intended application—industrial-grade enzymes may tolerate minor impurities, while pharmaceutical applications demand high purity (Zhu et al. 2024).

2.4 *Characterization of Enzymes*

Enzymes are biological catalysts made up of proteins that accelerate chemical reactions in living organisms. They are highly specific to the substrates they act upon, meaning each enzyme catalyzes only particular reactions. Enzyme characterization focuses on understanding their structure, including the active site, where the substrate binds and the reaction occurs. The enzyme's three-dimensional structure is crucial for its function, and even small changes in this structure can affect its efficiency and specificity. Enzymes also exhibit optimal conditions, such as temperature and pH, where they function best. Outside these conditions, enzymes can become denatured, losing their catalytic activity. Additionally, the study of enzyme kinetics helps determine their efficiency in catalyzing reactions. This involves measuring parameters such as the Michaelis constant (Km) and maximum reaction velocity (Vmax), which reflect the enzyme's affinity for the substrate and its catalytic efficiency. Enzyme activity can also be influenced by various factors like the presence of inhibitors, cofactors, and coenzymes, which can enhance or inhibit enzyme function. Characterizing enzymes is crucial for a variety of industries, including pharmaceuticals, biofuels, and food processing, as it enables the optimization of enzyme performance in different applications. Understanding these characteristics ensures that enzymes can be effectively utilized in biotechnological and industrial processes.Characterizing enzymes involves evaluating their physical, chemical, and kinetic properties to determine their suitability for specific applications. Key parameters include:

1. **Activity and Stability**: Enzyme activity is typically measured by the rate of substrate conversion under standard conditions. Stability assessments involve determining the enzyme's resistance to denaturation caused by temperature, pH, or chemical agents.
2. **Optimum Conditions**: Identifying the optimal pH and temperature for enzyme activity is crucial. For instance, thermostable enzymes, such as those from thermophilic microorganisms, are desirable for high-temperature industrial processes.
3. **Kinetics**: Enzyme kinetics, described by parameters like Km (Michaelis constant) and Vmax (maximum reaction velocity), provides insights into enzyme efficiency and substrate affinity.
4. **Structural Characterization**: Advanced techniques like X-ray crystallography and nuclear magnetic resonance (NMR) spectroscopy reveal the three-dimensional structure of enzymes, aiding in understanding their catalytic mechanisms and enabling rational design for improved performance.

Advances in Enzyme Production and Characterization

Modern advancements have revolutionized enzyme production and characterization. Genetic engineering enables the development of recombinant enzymes with enhanced activity, stability, and specificity. For example, site-directed mutagenesis allows precise modifications in enzyme structure to tailor its properties for particular applications. Additionally, immobilization technologies—where enzymes are fixed on solid supports—enhance enzyme reusability and process stability.

High-throughput screening methods, combined with bioinformatics tools, have accelerated the discovery of novel enzymes from metagenomic libraries. Machine learning algorithms further aid in predicting enzyme properties and guiding engineering efforts. These innovations, coupled with advancements in bioreactor design and process optimization, continue to expand the scope of enzymatic applications across industries.

1. **Types of Enzymatic Reactions**

Enzymatic reactions are classified according to the kind of chemical changes they catalyze. For enzymes to be used in biotechnology and industrial processes, it is crucial to comprehend their diverse forms. Numerous biochemical processes are supported by enzymes, which are extremely selective biocatalysts. The following are the primary categories of enzymatic reactions:

2. **Hydrolysis Reactions**

Hydrolysis processes break down substrates when water is supplied. The enzymes that catalyze these reactions are called hydrolases. Hydrolysis is a crucial stage in digestion and the breakdown of complex macromolecules into smaller parts. These are some examples of amylases, which are enzymes that hydrolyze starch into sugars.

Enzymes called proteases break down proteins into peptides and amino acids.

Glycerol and fatty acids are produced when fats and oils are hydrolyzed by enzymes called lipases.

3. **Oxidation-Reduction Reactions**

In oxidation-reduction (redox) reactions, molecules change their oxidation states by exchanging electrons. Enzymes called oxidoreductases facilitate these processes. Energy production and metabolism depend on these mechanisms. Examples include:

One enzyme that catalyzes the extraction of hydrogen atoms from substrates is alcohol dehydrogenase.

Enzymes called oxidases, like cytochrome oxidase, catalyze the transformation of electrons into oxygen.

4. **Transfer Reactions**

During transfer processes, functional groups are moved from one molecule to another. The enzymes that catalyze these reactions are called transferases. Many metabolic pathways rely on these mechanisms. Examples include:

The enzymes known as aminotransferases, such alanine transaminase, are responsible for transferring amino groups from amino acids to alpha-keto acids.

Phosphate groups are transferred from ATP to substrates by kinases, which are enzymes like hexokinase.

5. **Lyase Reactions**

Chemical groups are added to or taken away from double bonds in lyase reactions, forming new bonds without oxidation or hydrolysis. The enzymes that catalyze these reactions are known as lyases. Enzymes called decarboxylases take carboxyl groups out of substrates, often releasing carbon dioxide in the process. One example is pyruvate decarboxylase.

Enzymes known as aldolases catalyze the breakdown of carbon-carbon bonds in carbohydrates.

6. **Isomerization Reactions**

During isomerization reactions, atoms within a molecule can rearrange, producing isomers in the process. These reactions are catalyzed by enzymes called isomerases. These reactions are necessary for metabolic activities that transform substrates into different forms. Examples include:

One stereoisomer is changed into another by enzymes called racemases, such as amino acid racemase.

Mutases are enzymes that alter the functional groups in molecules, such phosphoglycerate mutase.

7. **Ligase Reactions**

Ligase reactions include the joining of two substrates and the simultaneous hydrolysis of ATP or another high-energy molecule. The enzymes that catalyze these processes are called ligases and synthetases. Biosynthesis pathways rely on these mechanisms. Here are a few examples:

During replication and repair, DNA fragments are joined by an enzyme called DNA ligase.

Acetyl-CoA synthetase is the enzyme that catalyzes the transformation of CoA and acetate into acetyl-CoA (Fig. 5.2).

8. **Enzyme Kinetics**

Enzyme kinetics is the study of enzymatic reaction rates and the factors influencing them. Understanding enzyme kinetics is essential for developing drugs,

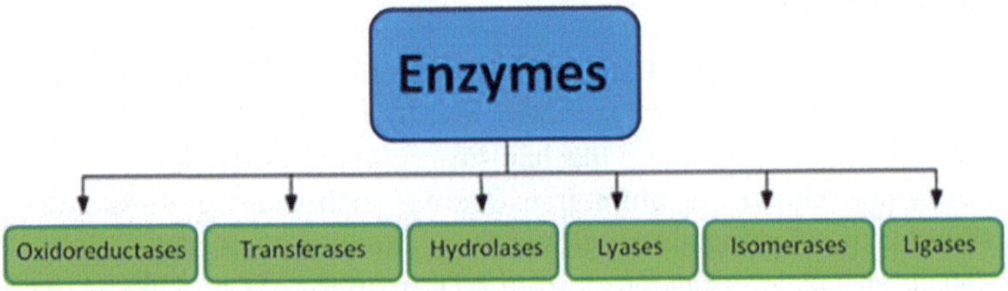

Fig. 5.2 Enzymes can be defined as biological polymers that catalyze biochemical reactions. (Wulff, 2002)

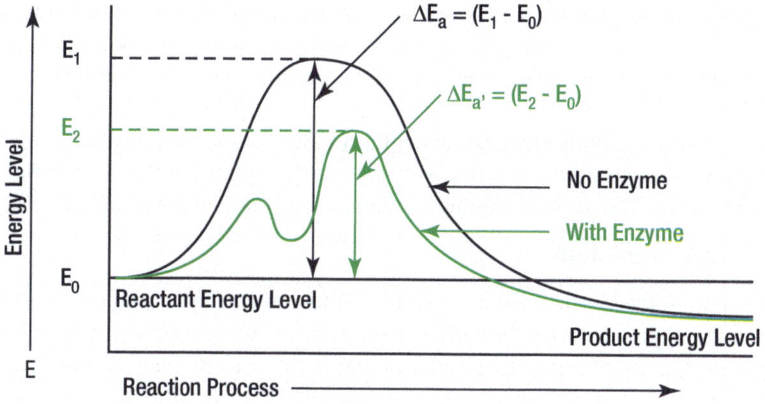

Fig. 5.3 Enzyme kinetics: energy levels. (Bisswanger, 2017, 21)

optimizing industrial processes, and elucidating enzyme mechanisms. The primary model used to explain enzyme kinetics is the Michaelis-Menten model, which explains the relationship between substrate concentration and reaction velocity. This section will explain the fundamental concepts of enzyme kinetics, including key terms, the Michaelis-Menten equation, factors affecting enzyme activity, and enzyme control (Fig. 5.3).

9. **Michaelis-Menten Kinetics**

The Michaelis-Menten model explains the connection between substrate concentration and reaction velocity. According to this theory, the enzyme and substrate combine to form an enzyme-substrate complex, which then changes into a product and releases the free enzyme. The key equation is:

10. Factors Affecting Enzyme Activity

- Substrate Concentration: Increasing substrate concentration often accelerates the process until the enzyme achieves saturation. The pace of reaction then levels out.

Temperature: Enzyme activity typically increases up to an ideal temperature; above that, denaturation of the enzyme may occur, which would result in decreased activity.
pH: Each enzyme has a preferred pH range. Deviations from this range could affect the ionization of active site residues and alter the enzyme's activity.

- Enzyme Concentration: Increasing the enzyme concentration typically results in a faster reaction when there is an excess of substrate.

11. **Enzyme Inhibition**

Enzyme activity can be inhibited by a variety of substances. Two fundamental types of inhibition exist:

- Competitive Inhibition: The active site is the target of a binding competition between an inhibitor and substrate. By raising the substrate concentration, this inhibition can be lessened. This type of inhibition raises Km but has no effect on Vmax.
- Not competitive Inhibition: By attaching to an enzyme at a position other than the active site, an inhibitor reduces its activity regardless of the concentration of the substrate. This type of inhibition reduces V_{max} without affecting K_m.

12. **Enzyme Regulation**

To maintain metabolic balance, a variety of techniques are commonly employed to regulate enzyme activity. Important regulatory mechanisms include the following: Allostery: There are active and inactive forms of enzymes. Conformational changes brought on by allosteric activators and inhibitors binding to sites other than the active site affect the activity of the enzyme.

Feedback Inhibition: An enzyme that was involved earlier in the pathway may be inhibited by the product of a metabolic process to prevent overproduction.

Covalent alteration: One type of covalent change that can either activate or deactivate enzymes is phosphorylation.

13. **Enzyme Sources**

A diverse range of organisms, including bacteria, plants, and mammals, produce enzymes, which are biocatalysts. Each source offers unique advantages and applications based on the characteristics of the specific enzyme, such as stability, activity, and substrate specificity. This section will cover the three primary sources of enzymes—microbial, plant, and animal—as well as their significance in various industrial contexts.

1. Microbial Enzymes

Microbial enzymes, produced by bacteria, yeast, and fungus, are widely used in a wide range of industrial processes due to their high catalytic efficiency, specificity, and flexibility to a variety of environmental situations.

Bacteria: Numerous enzymes, such as lipases, amylases, and proteases, are produced by certain bacteria, such as *Bacillus* and *Escherichia coli*. Their enzymes are commonly used in industrial settings because of their adaptability and ability to thrive under challenging conditions, such as high temperatures or pH fluctuations.

Yeast: Yeasts like *Saccharomyces cerevisiae* are commonly used in the production of food and beverages. They produce enzymes like invertase, which catalyzes the conversion of sucrose into glucose and fructose and is essential for baking and brewing.

- fungus: Filamentous fungi, such as *Aspergillus* and *Penicillium*, generate a variety of enzymes, including cellulases and kininases, which are used in the breakdown of plant materials and bioconversion processes. Fungal enzymes are very helpful for biofuel production and waste treatment.

2. Plant Enzymes

Plant enzymes are made from a variety of plant parts, including seeds, fruits, leaves, and roots. They play crucial roles in plant metabolism and are increasingly well-known for their industrial applications.

- Sources: Common plant sources of enzymes include papaya (papain), pineapple (bromelain), and figs (ficin). These enzymes are commonly used in food processing, meat tenderization, brewing, and juice clarity.

Applications: In certain applications, plant enzymes are chosen over microbial enzymes due to their greater specialization and perceived "naturalness." Dietary supplements for digestive health also contain them.

3. Animal Enzymes

Animal-derived enzymes are typically found in animal tissues, organs, or digestive tracts. Throughout history, they have been used in a wide range of applications due to their effectiveness and distinct activity.

Sources: Common sources include pancreatin, which is derived from the pig pancreas and contains lipases, amylases, and proteases, and rennin (sometimes called chymosin), which is taken from the stomachs of calves and used to produce cheese.

Applications: Animal enzymes are used in food processing, pharmaceutical formulations, and diagnostic instruments. However, their use has decreased because to concerns about animal welfare and the availability of plant and microbial replacements.

14. **Applications of Enzymatic Conversion**

Enzymatic conversion is crucial in a wide range of industries because it accelerates biochemical processes by utilizing the unique properties of enzymes. This technology is quite popular because of its efficiency, specificity, and environmental friendliness as compared to traditional chemical processes. The following is a list of the specific applications of enzymatic conversion in different sectors.

1. Food Industry

In the food industry, enzymatic conversion is frequently employed to enhance flavors, improve textures, and increase nutritional content. Enzymes, such as cellulases, amylases, and proteases, have a variety of applications. For instance, amylases aid in fermentation by turning starches into sugars during baking and brewing. While proteases break down proteins to tenderize meat, cellulases are used to clarify fruit juices and improve the texture of processed foods. These enzymatic processes not only improve product quality but also increase production efficiency.

2. Beverage Production

In the fermentation process used to produce beverages, enzymatic conversion is crucial. Yeast enzymes are essential for fermenting carbohydrates into alcohol in the production of beer, wine, and spirits. Enzymes such as pectinase break down the pectin in fruits, resulting in clearer juices and wines. Additionally, glucanases and other enzymes enhance the extraction of tastes and colors from grains and fruits, improving the overall quality of the final products. This application demonstrates how enzymatic conversion can enhance the effectiveness and quality of beverage production.

3. Dairy Industry

The dairy business uses enzymatic conversion to produce a range of products, including cheese, yogurt, and whey protein. The rennet enzyme, which is taken from the stomachs of young calves, is crucial for coagulating milk and separating curds from whey in the cheese-making process. Other enzymes, like lactase, hydrolyze lactose in milk so that lactose-intolerant individuals can digest it. Additionally, proteases are utilized to improve the texture and flavor of cheese as it ripens. These enzymatic processes significantly raise the caliber of dairy products and increase their attractiveness to consumers.

4. Biofuel Production

Enzymatic conversion is essential to the production of biofuels, particularly bioethanol and biodiesel. During the production of bioethanol, lignocellulosic biomass—such as wood and agricultural waste—is converted into fermentable sugars by enzymes including cellulases and amylases. Yeast subsequently ferments these sugars to create ethanol. Through the reduction of greenhouse gas emissions and dependence on non-renewable energy sources, this strategy offers a sustainable alternative to fossil fuels. Moreover, the transesterification of fats and oils is catalyzed by lipases, providing a cleaner and more efficient way to produce biodiesel.

5. Pharmaceutical Industry

Enzymatic conversion is widely used in the pharmaceutical industry to produce and break down medications. The biotransformation mechanisms that yield active pharmaceutical ingredients (APIs) depend on enzymes. For example, cytochrome P450 enzymes are used to add functional groups to pharmaceutical substances in order to boost their therapeutic efficacy. Gaucher's disease and other hereditary disorders where specific enzymes are deficient are also treated using enzyme replacement therapies. This application demonstrates how enzymatic conversion can improve drug development and personalized treatment.

6. Bioremediation

Enzymatic conversion is also crucial in bioremediation, which uses enzymes to degrade poisons and contaminants in the environment. Laccases and peroxidases, for instance, can degrade lignin and other complex organic materials found in contaminated soils and water. In wastewater treatment, urease and other enzymes aid in the breakdown of urea, reducing eutrophication and nitrogen levels. This application illustrates how enzymatic conversion can be applied to promote sustainability and address environmental problems.

7. Textile Industry

The textile industry uses enzymatic conversion to clean and treat textiles. Enzymes such as cellulases, pectinases, and proteases are utilized in processes like biopolishing, which enhances the appearance and texture of textiles by removing excess fibers and impurities. Denim can also be polished using enzymes to give it a worn-out look without the need of harsh chemicals. Enzymatic conversion in textiles improves product quality and promotes sustainable practices by reducing the environmental impact of traditional textile processing methods.

8. Animal Feed Production

Enzymatic conversion is increasingly being employed in animal feed production to increase nutritional availability and digestibility. Enzymes such as phytases and xylanases are added to animal diets to improve nutrient absorption and degrade antinutritional components. For example, when phytases hydrolyze phytic acid, phosphorus is liberated and more accessible to monogastric animals such as pigs and chickens. This application facilitates better animal health and development rates, reduced feed expenses, and more efficient feed utilization.

9. Cosmetics and Personal Care

The cosmetics and personal care industries use enzymatic conversion to create innovative products that meet a variety of consumer needs. Examples of enzymes used in skin exfoliation and cleansing formulations to give gentle, yet effective skincare treatments are lipases and proteases. In order to promote healthy hair and scalp, hair care products also contain enzymes. Enzymatic conversion in cosmetics satisfies consumer demands for natural and environmentally safe components while simultaneously enhancing product performance.

10. Agricultural Biotechnology

Enzymatic conversion is a common technique in agricultural biotechnology, particularly in the production of biopesticides and biofertilizers. Enzymes aid in the growth and increased production of plants by improving the availability of nutrients in the soil. For example, nitrogen-fixing bacteria create enzymes that change atmospheric nitrogen into a form that plants can absorb. In order to target specific pests with the least amount of harm to beneficial organisms, enzymes are also utilized in biopesticides. The potential of enzymatic conversion to advance sustainable agriculture and enhance food security is demonstrated by this application.

15. Advancements in Enzymatic Conversion Technologies

Thanks to advancements in genetic engineering, biotechnology, and process optimization, enzymatic conversion has made considerable strides throughout time. These advancements have improved the efficiency, specificity, and industry-wide applicability of enzymes. The following noteworthy advancements in enzymatic conversion technologies will have an impact on this topic in the future.

1. Enzyme Engineering

Enhancing the properties of enzymes using techniques like directed evolution and rational design has been the aim of recent advancements in enzyme engineering. Directed evolution allows scientists to create enzyme variants with enhanced stability, specificity, or activity by mimicking natural selection in a laboratory setting. Rational design includes the use of computational techniques to predict the impact of mutations on enzyme activity. Enzymes with improved performance and affordability that are appropriate for certain industrial applications have been created using these techniques.

2. Metagenomics

Without the need for laboratory organism cultivation, novel enzymes can be identified from ambient materials using a powerful technique called metagenomics. By sequencing DNA from a range of microbial communities, scientists can identify genes encoding unique enzymes that may have superior properties than known enzymes. This process has expanded the variety of enzymes available, leading to advancements in biocatalysis, bioremediation, and biofuel generation, among other areas. Utilizing the enormous diversity seen in nature opens up new avenues for enzyme discovery and use.

3. Immobilization Techniques

Immobilization onto solid substrates has become more and more common as a way to increase the stability, reusability, and convenience of separating enzymes from reaction mixtures. Advanced immobilization methods such as covalent bonding, adsorption, and encapsulation allow for the effective fixing of enzymes while maintaining their activity. Improvements in support materials like magnetic nanoparticles and nanofibers have significantly improved the functionality of immobilized enzymes in industrial processes. These advancements contribute to the cost-effectiveness and efficiency of enzymatic reactions.

4. Microbial Fermentation

The latest developments in microbial fermentation technologies have optimized the production of enzymes on a larger scale. Improvements in fermentation technology, such as fed-batch and continuous fermentation systems, enable the manufacture of enzymes from microbial sources. Utilizing genetically modified organisms (GMOs) with enhanced metabolic pathways can also lead to higher enzyme production. Improvements in fermentation technology have made enzyme synthesis more economically feasible and made enzymatic conversion processes more competitive with traditional chemical methods.

5. Bioreactor Design

Improvements in bioreactor design have significantly changed the efficiency of enzyme conversion processes. Modern bioreactors are equipped with sophisticated monitoring and control systems that enable real-time parameter tuning, including substrate concentration, pH, and temperature. Microfluidics and lab-on-a-chip technologies provide high-throughput screening of enzyme stability and activity under a variety of conditions. Enzymatic processes are now more scalable thanks to advancements in bioreactor technology, opening the door to their application in industrial settings.

6. Systems Biology and Omics Technologies

The application of systems biology and omics technologies (genomics, proteomics, and metabolomics) has enabled a better understanding of the functions and interactions of enzymes in biological systems. By looking at the complete set of genes, proteins, and metabolites, researchers can identify the metabolic pathways and regulatory mechanisms involved in enzymatic conversion. This thorough understanding enables the logical construction of enzyme mixes for specific applications, improving the overall efficiency and performance of biotechnological processes.

7. Artificial Intelligence and Machine Learning

The application of machine learning (ML) and artificial intelligence (AI) to enzymatic conversion research is revolutionizing enzyme discovery and improvement. The finding of suitable candidates for industrial applications is made easier by AI algorithms' abilities to examine vast datasets and forecast enzyme activity, stability, and substrate specificity. Additionally, machine learning models can improve the efficiency of enzymatic conversions by refining reaction conditions. These technological advancements enable researchers to make data-driven decisions, which expedites the discovery and use of enzymes.

8. Green Chemistry Approaches

The ideas of green chemistry, which aims to lessen hazardous substances and their adverse environmental effects, are closely tied to advancements in enzymatic conversion. Energy usage is often reduced and harsh chemicals are often avoided because enzymes work best in moderate settings. Since the use of biocatalysis in synthesis and processing is in line with sustainable practices, enzymatic conversion is an essential component in the development of environmentally friendly industrial processes. Because sustainability is emphasized, enzyme technology is always changing.

9. Integration with Synthetic Biology

The integration of enzymatic conversion technology and synthetic biology is propelling biomanufacturing to new heights. Synthetic biology allows for the creation of modified bacteria that can use enzyme pathways to synthesize useful compounds by conceiving and constructing new biological systems. This approach could lead to the development of eco-friendly methods for making biofuels, pharmaceuticals, and specialized chemicals. When combined, enzymatic conversion and synthetic biology may be able to resolve challenging issues in production and resource management.

10. Personalized Medicine

Advances in enzymatic conversion are paving the way for customized medicine, especially with the introduction of enzyme-based therapies. Enzymes designed to target specific metabolic pathways or hereditary illnesses can be used to provide individualized therapy for each patient. For instance, enzyme replacement therapy is being used to treat lysosomal storage disorders and other conditions brought on by a lack of certain enzymes. The ability to customize enzyme therapy is a significant step in addressing each person's unique health needs.

16. Challenges in Enzymatic Conversion

Enzymatic conversion is widely used in many industries due to its efficiency and specificity, however several barriers stand in the way of its widespread application and advancement. Understanding these challenges is essential to the development of enzyme technologies and their increased utility in industrial processes. The following is a list of the primary barriers to enzymatic conversion.

1. Enzyme Stability and Activity

One of the main challenges in enzymatic conversion is maintaining enzyme stability and activity under working conditions. Temperature, pH, and the presence of inhibitors or denaturing chemicals can all affect how sensitive an enzyme is. Denaturation-induced activity loss can significantly reduce the efficiency of enzyme activities. Therefore, it is essential to identify and enhance the conditions that sustain enzyme stability throughout time. Despite being a recurring challenge, this issue can be mitigated through the development of more resilient enzymes or the application of immobilization techniques.

2. Substrate Specificity and Availability

Since enzymatic conversion frequently depends on certain substrates, fluctuations in substrate supply may restrict the process's effectiveness. For instance, high-purity substrates, which can be expensive and time-consuming to create, are necessary for many industrial applications. Furthermore, because of their selectivity, certain enzymes may have limited substrate ranges or be unable to convert specific substrates well. Developing enzymes with higher specificity or simplifying processes to function with other substrates can get over these limitations, but the expense of research and development for these solutions is high.

3. Cost-Effectiveness

The economic feasibility of enzymatic conversion technologies is a significant obstacle when contrasted with traditional chemical methods. The cost of generating, purifying, and stabilizing enzymes can have an impact on the total cost of enzymatic operations. Additionally, there are numerous barriers that can affect cost effectiveness when transferring from laboratory-scale to industrial-scale production, making the scalability of these processes much more challenging. Lowering the cost of enzyme production and improving process economics through integration and optimization with existing technologies are necessary for wider use.

4. Reaction Kinetics and Process Optimization

The complex dynamics of enzymatic activities may not always allow for high conversion rates. Variables that can significantly affect process efficiency include reaction duration, substrate concentration, and enzyme concentration. Optimizing these parameters to get the desired reaction speeds and yields can be challenging and sometimes requires extensive testing. The use of advanced modeling techniques and high-throughput screening technologies can help optimize processes, even while the complexity of enzyme reactions remains a hurdle.

5. Integration with Existing Processes

It may be challenging to include enzymatic conversion into existing industrial processes due to compatibility issues. Since chemical catalysts constitute the foundation of many traditional processes, switching to enzymes may require considerable adjustments. This integration can include staff management training for enzymatic processes, equipment improvements, and changes to operating protocols. These barriers must be eliminated in order to include enzymatic conversion into existing workflows and fully appreciate its potential benefits.

6. Regulatory and Safety Concerns

There are stringent and intricate regulatory systems governing the use of enzymes in the food, pharmaceutical, and other industries. Enzyme-based product commercialization hinges on compliance with safety regulations and laws. Additionally, the potential for allergenicity and unpleasant reactions must be assessed, especially in food applications. Overcoming regulatory barriers and conducting thorough safety evaluations are necessary for the development and approval of enzymatic procedures.

7. Environmental Factors

Environmental influences can have a big impact on enzymatic processes. A number of variables, including temperature, pH, ionic strength, and the presence of contaminants, can alter the stability and activity of enzymes. Furthermore, the use of enzymes in bioremediation and waste treatment processes may be influenced by the characteristics of the waste stream, including its composition and pollutant level. Addressing these environmental limits requires careful consideration of the operational conditions and potential enzymatic process changes.

8. Enzyme Deactivation

Enzyme deactivation throughout the reaction could significantly reduce yield and overall efficiency. Substrate inhibition, product inhibition, and enzyme degradation are some of the variables that might lead to a decline in activity over time. Understanding the reasons behind enzyme deactivation and developing mitigating strategies, including reaction condition adjustment or the use of enzyme stabilizers, are necessary to improve the reliability of enzymatic conversion processes.

9. Public Perception and Acceptance

Public acceptance and perception of enzymatic processes can be challenging, particularly in the food and pharmaceutical industries. Concerns over the safety of enzyme applications, genetically modified organisms (GMOs), and enzyme sources may lead to consumer resistance. Public education regarding the benefits and safety of enzymatic conversion technologies is necessary to promote acceptance and usage in a variety of industries.

10. Research and Development Investment

The advancement of enzymatic conversion technology requires significant investment in research and development. Because many companies prioritize short-term benefits over long-term gains, funding for enzyme research may be inadequate. Continuous investment is required to create new enzymes, improve existing processes, and look into creative uses in order to advance technical advancements. Fostering collaboration between academia, industry, and government can support the development of enzymatic technologies and encourage research.

17. Future Trends in Enzymatic Conversion

Future advances in enzymatic conversion are expected to focus on improving scalability, sustainability, and efficiency as a result of advancements in synthetic biology and biotechnology. Researchers are creating specialized enzymes with increased stability, specificity, and activity through the vital field of enzyme engineering. This could lead to more efficient processes for making bioplastics, drugs, and biofuels. Because computational technologies like artificial intelligence and machine learning shorten the duration and lower the cost of experimental procedures, they are becoming more and more significant in enzyme discovery and optimization. Another noteworthy trend is the combination of enzymatic processes with renewable and environmentally friendly materials. Enzymatic conversions provide a more environmentally friendly option to conventional chemical processes as corporations attempt to minimize their carbon footprint. Enzymes can aid in the conversion of plant biomass, agricultural waste, and even algae into sustainable fuel sources, making this trend especially pertinent to the creation of biofuel. It is anticipated that the creation of new enzymes capable of more effectively breaking down resistant substances, such as lignocellulose, will increase the cost-effectiveness and scalability of biofuel production, supporting the circular bioeconomy. Lastly, bioprocess optimization is becoming increasingly relevant as firms seek to increase enzymatic conversions for commercial usage. Continuous research into immobilized enzymes, enzyme cascades, and multi-enzyme systems allows for more efficient industrial processes with reduced operational costs. These advances are critical for expanding the applications of enzymatic conversions in industries like food, pharmaceuticals, and waste management. Moreover, the use of enzymes in bioelectrochemical systems, where they facilitate reactions at electrodes, is a growing area with potential for converting waste into fuel or electricity, further contributing to sustainable energy solutions.

3 Conclusion

Enzymatic conversion is a highly efficient and environmentally friendly process that utilizes enzymes as catalysts to transform substrates into desired products.This process is widely used in various industries, including food, pharmaceuticals, biofuels, and waste treatment, due to its specificity, milder reaction conditions,and reduced production of harmful by-products compared to traditional chemical methods. The ability of enzymes to work under mild conditions such as lowtemperatures and neutral pH further enhances their appeal for sustainable production processes. Overall, enzymatic conversion offers a promising alternative toconventional methods, contributing to more sustainable and cost-effective industrial processes.

References

Bisswanger, H. (2017). *Enzyme kinetics: principles and methods*. John Wiley & Sons.

Liang, J., Zhang, R., Chang, J., Chen, L., Nabi, M., Zhang, H., Zhang, G., Zhang, P. (2024). Rumen microbes, enzymes, metabolisms, and application in lignocellulosic waste conversion-A comprehensive review. *Biotechnology Advances*. 1,71:108308.

Marpani, F., Pinelo, M., & Meyer, A. S. (2017). Enzymatic conversion of CO2 to CH3OH via reverse dehydrogenase cascade biocatalysis: Quantitative comparison of efficiencies of immobilized enzyme systems. *Biochemical Engineering Journal, 127*, 217–228.

Wulff, G. (2002). Enzyme-like catalysis by molecularly imprinted polymers. *Chemical Reviews, 102*(1), 1–28.

Zhu, S,Y., Liu, S,C., Zhang, C,X., Xin, X., Liu, Z,H., Zhang, L,J., Li, B,Z., Yuan, Y,J. (2024). Pathway and enzyme engineering for the bioconversion of lignin derivatives into homoeriodictyol in Saccharomyces cerevisiae. *Green Chemistry*. *26*(9):5260–72.

Chapter 6
Genetic Modification and Biotechnology

Abdul Sattar Jatoi, Arshad Iqbal, and Bazil Owais Nasir

1 Introduction

Genetic modification (GM) is the direct manipulation of an organism's genetic material to change its characteristics. By employing complex techniques to introduce, remove, or alter genes, scientists can create organisms with desired traits. Because genetic modification can accurately alter DNA, it differs from traditional breeding, which relies on natural processes and may take generations to produce comparable effects. Biotechnology encompasses a broader variety of technologies that use biological systems, organisms, or derivatives to produce products or solve problems. Genetic modification, one of the core tools of biotechnology, provides advancements in a variety of domains, including agriculture, medicine, industry, and environmental management. The potential for genetic alteration to address global problems including disease prevention, food security, and environmental sustainability makes it significant. In agriculture, genetically modified organisms (GMOs) have been developed to boost crop yields, enhance pest resistance, and strengthen crops' ability to withstand harsh weather conditions. In areas affected by food shortages and climate change, this has proved very helpful. Genetic engineering has transformed the manufacturing of medications such as insulin, vaccinations, and monoclonal antibodies in the medical field. Additionally, it enabled gene therapy, which fixes genetic flaws in people with inherited illnesses. Furthermore, industrial biotechnology's use of genetic alteration.

Despite its immense potential, genetic modification is a subject of ethical debate and regulatory scrutiny.. Concerns regarding the long-term ecological effects of genetically modified organisms (GMOs), the alteration of human DNA, and the

A. S. Jatoi (✉) · A. Iqbal · B. O. Nasir
Department of Chemical Engineering, Dawood University of Engineering and Technology, Karachi, Pakistan
e-mail: abdul.sattar@duet.edu.pk

A. S. Jatoi (ed.), *Integrated Food and Bioprocessing*,
https://doi.org/10.1007/978-3-031-87209-9_6

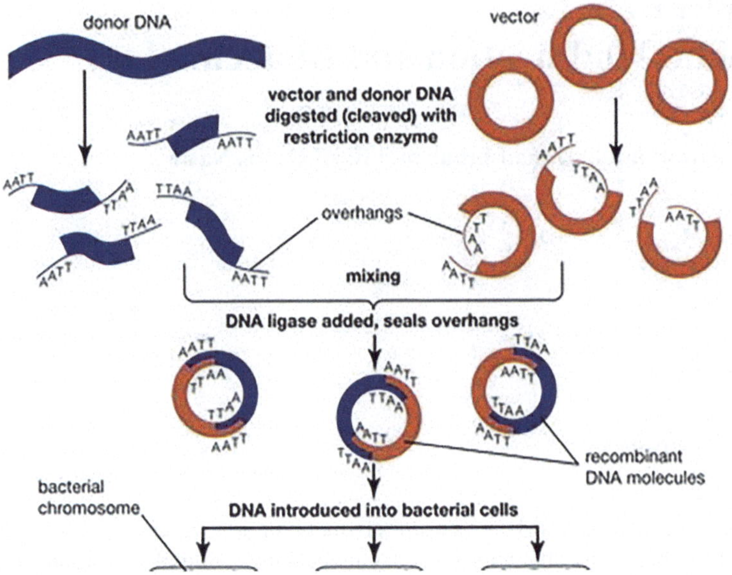

Fig. 6.1 Carnegie endowment for international peace—the blessing and curse of biotechnology. (Langer & Sharma, 2020)

unforeseen consequences of gene editing have raised questions about the responsible use of this technology. Regulatory bodies throughout the world have established frameworks to ensure that genetically modified products are safe for the environment and human health.. Whether or not GM technologies are accepted and used also heavily depends on the public's perception, which is shaped by cultural, ethical, and scientific issues. As biotechnology advances, finding a balance between innovation, moral responsibility, and safety will remain a top challenge for academics, policymakers, and the general public (Fig. 6.1).

2 Basic Concepts in Genetic Modification

In order to change an organism's traits or capacities, genetic alteration entails changing its genetic material, usually its DNA. Nucleotide sequences that encode genes make up DNA, the molecule that houses genetic information. The physical and biological characteristics of an organism are determined by these genes. By precisely adding, removing, or altering genes through genetic modification, scientists can produce novel traits, such as disease resistance in plants or the creation of therapeutic proteins in bacteria. Transgenic animals can be produced by introducing genes from one species into another since the genetic code is universal.

Over time, genetic modification techniques have seen significant change. One of the most popular methods is recombinant DNA technology, which creates novel

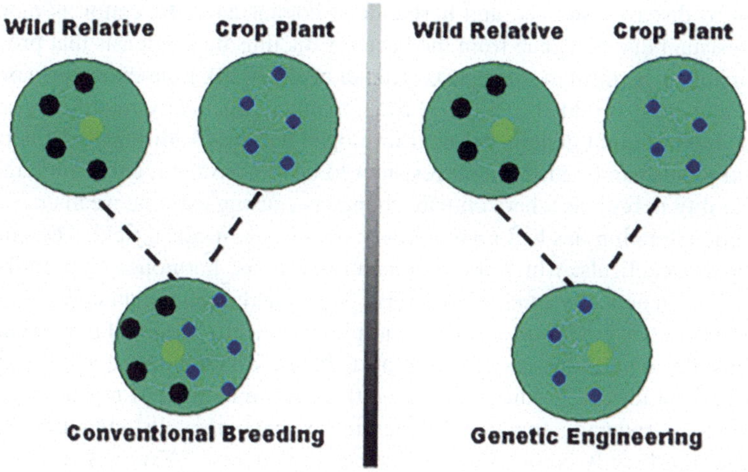

Fig. 6.2 Genetic engineering and GM crops. (Kumar et al., 2020)

genetic sequences by combining DNA molecules from various sources. This recombinant DNA is then inserted into a host organism, usually via a vector like a virus or plasmid, to facilitate gene transfer. Another innovative tool is the precise genome-editing method CRISPR-Cas9, which can cut and alter specific DNA sequences. Unlike earlier methods that could only add or delete genes, CRISPR allows for very focused editing, making it a powerful tool for gene correction in medical therapies and the development of genetically modified crops and animals.

Various gene transfer methods are employed, depending on the type of organism and the intended modification. Through transformation, in which the organism takes up DNA from its environment, genes can be inserted into microorganisms like bacteria. More advanced methods are used to introduce new genes for plants and animals, such as gene gun technology or Agrobacterium-mediated transfer. In animals, viral vectors are commonly employed to introduce genes into cells.. Researchers can use these methods to produce genetically modified organisms (GMOs) that have desired traits, such increased nutritional value, crop resistance to insects, or even the ability to produce therapeutic drugs. But as new tools and techniques are created, these processes continue to get more precise and efficient, pushing the boundaries of genetic modification (Fig. 6.2).

3 Applications of Genetic Modification

The use of genetic engineering to create genetically modified crops (GMOs) to address problems such as pest infestation, poor soil quality, and climate change has fundamentally altered agriculture. Farmers may boost agricultural output and significantly reduce their reliance on chemical pesticides by growing crops that are resistant to diseases, insects, and herbicides.. For instance, Bt cotton is naturally insect-resistant due to a gene from the bacteria Bacillus thuringiensis that produces a protein that is harmful to some pests (Kumar et al., 2020). Beta-carotene, a precursor of vitamin A, is added to Golden Rice, another genetically modified grain, to assist address vitamin deficits in underdeveloped nations. Additionally, genetically modified plants can be made more resistant to heat or drought, enhancing agricultural viability in regions where climate change is causing extreme weather.

Genetic alteration has had revolutionary uses in the medical field. The creation of biopharmaceuticals, which are therapeutic proteins, hormones, and antibodies produced by genetically modified bacteria, yeast, and mammalian cells, is one of the most important contributions. For example, genetically modified *E. coli* bacteria create insulin, a hormone needed to treat diabetes, which eliminates the need for animal-derived insulin. Additionally, gene therapy—a method of repairing genetic diseases by introducing functional genes into a patient's cells—requires genetic alteration. By directly editing defective genes in patients' DNA, new technologies like CRISPR-Cas9 are being investigated for the treatment of diseases like sickle cell anemia, cystic fibrosis, and some types of cancer. This holds out promise for the cure of genetic disorders that were previously incurable.

Genetic modification is also advantageous for industrial biotechnology and the environment. Using genetically engineered organisms to clean up environmental contaminants like heavy metals or oil spills is known as bioremediation. By converting poisonous compounds into less dangerous products, genetically modified microorganisms can support environmental conservation initiatives. Genetically modified bacteria and algae are being developed to effectively manufacture biofuels from biomass in biofuel production, providing a sustainable substitute for fossil fuels. Cleaner, more sustainable industrial processes are made possible by the use of genetically engineered microorganisms in the production of biodegradable polymers, detergent enzymes, and other biochemicals. These developments in genetic manipulation are assisting industry in switching to more environmentally friendly options and fostering the expansion of bio-based economies (Fig. 6.3).

4 Advanced Techniques in Biotechnology

One of the most significant advancements in biotechnology is the development of CRISPR-Cas9 and other genome-editing tools. By employing CRISPR (Clustered Regularly Interspaced Short Palindromic Repeats) to edit particular DNA areas, scientists can introduce mutations or correct genetic defects with unprecedented precision. The Cas9 enzyme, which functions similarly to molecular scissors, makes specific cuts in DNA to allow for the addition or removal of genetic material. The method has revolutionized genetic research and therapy by enabling targeted therapies for inherited diseases such as muscular dystrophy and sickle cell anemia. CRISPR is being used to improve traits including growth rate, disease resistance,

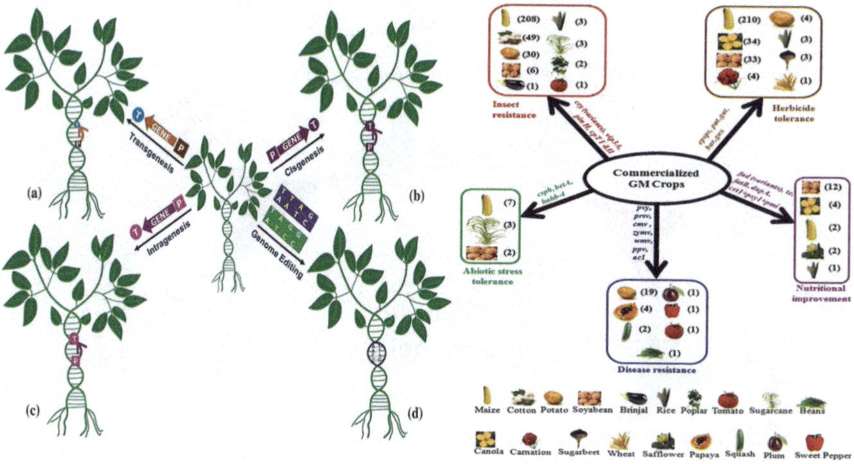

Fig. 6.3 Application genetic modification (GM) in food crops. (Kumar et al., 2020)

and nutritional value in animals and agricultural products in addition to its medical applications.

The field of synthetic biologyis another new field that focuses on creating new biological systems, tools, and components or repurposing biological systems for good. Scientists can build whole new life forms or alter organisms to carry out specific tasks, such making drugs, biofuels, or biodegradable materials, by combining concepts from biology and engineering. An important development in synthetic biology is the creation of synthetic genomes, in which scientists have reconstructed entire bacterial genomes from scratch and inserted them into cells. This method has very interesting applications in bio-manufacturing, where microbes can be altered to produce chemicals, drugs, or even food more effectively and sustainably.

Gene drives are another cutting-edge method that has a lot of potential in biotechnology, especially for managing pest and disease vector populations. Gene drives are genetic mechanisms that successfully disseminate genetic alterations throughout a population by increasing the possibility that a specific gene will be passed on to the following generation.. This approach is being researched to prevent the spread of diseases like malaria by genetically modifying mosquitoes to prevent them from transmitting the parasite. Gene drives offer a new approach to population control and may be used to control alien species or agricultural pests. However, due to potential ecological issues with releasing gene drive organisms into the wild, this technology is being closely examined by scientists and ethical experts (Fig. 6.4).

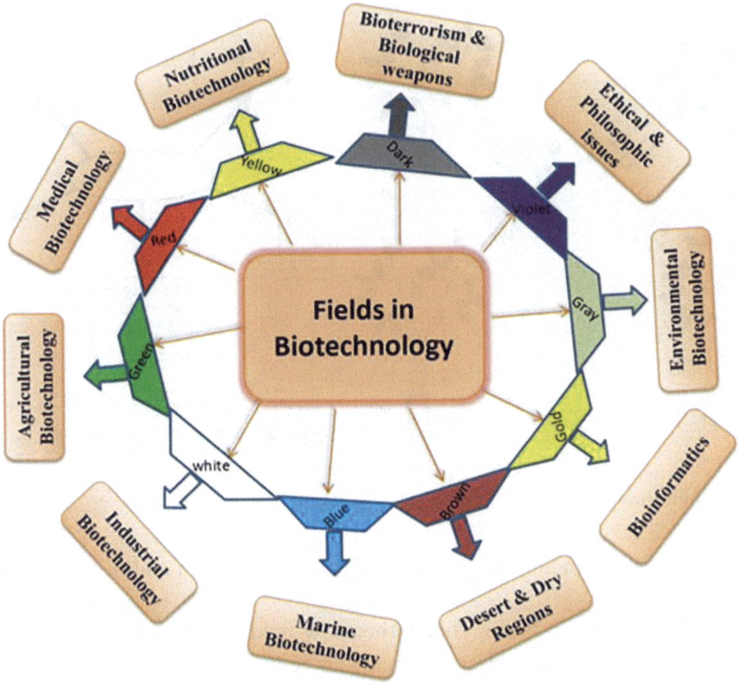

Fig. 6.4 Biotechnology courses—master the science behind life-saving innovations. (Cockburn & Stern, 2010)

5 Regulatory and Ethical Considerations

The rapid advancement of genetic modification and biotechnology has led to the establishment of several regulatory frameworks to guarantee the safety of genetically modified organisms (GMOs) and biotechnological products. Prior to GMOs being approved for use in commerce, In order to ascertain their safety, efficacy, and environmental impact, regulatory bodies such as the U.S. Food and Drug Administration (FDA), the European Food Safety Authority (EFSA), and the World Health Organization (WHO) are essential. These assessments often involve extensive testing and analysis of any risks associated with the introduction of genetically modified organisms (GMOs) into the environment or food supply, including unanticipated effects on biodiversity, human health, and ecosystems. Distinct countries have very distinct regulatory systems; some have more stringent laws than others, which often reflects various societal and cultural attitudes on biotechnology.

Genetic modification presents several ethical issues, from social justice issues to moral repercussions, all of which are equally important. The potential for genetic engineering to alter human germlines raises serious ethical questions regarding the alteration of human DNA and the long-term consequences of such modifications.

Concerns about consent are also very important, particularly in gene therapy as modifications may affect future generations. In addition, the commercialization of living things presents moral dilemmas about the ownership and patenting of genetic components, which may limit access to necessary technology and further the gap between wealthy and developing nations. Since these ethical challenges affect legislative decisions and the adoption of biotechnological developments, public participation and discussion are necessary to address them.

Finding a balance between inventiveness and moral responsibility is crucial in the fields of genetic modification and biotechnology. Ensuring that ethical considerations are integrated into the development and use of biotechnological products helps foster responsible innovation and public trust. It is necessary to include stakeholders, including scientists, ethicists, lawmakers, and the general public, in the development of open frameworks that address ethical challenges and progress knowledge. To navigate the complex landscape of genetic modification and biotechnology and ensure that their benefits are realized while reducing risks and ethical dilemmas, As the field evolves, ongoing debate and adaptable regulatory approaches will be essential.

6 Challenges and Future Prospects

One of the primary issues in genetic modification and biotechnology is ensuring the precision and safety of gene-editing techniques. Despite the fact that methods like CRISPR-Cas9 have greatly improved precision, Off-target effects, or inadvertent changes to DNA, are still an issue, especially in medical applications where errors could lead to undesirable genetic abnormalities or negative side effects. Delivering the altered genes to the appropriate cells in the patient's body in a safe and effective way is one of the main hurdles in gene therapy. Furthermore, it can be difficult to keep the changes stable over time, particularly in therapeutic applications where unanticipated consequences might manifest years after treatment. These challenges necessitate ongoing improvements in editing techniques and extensive testing prior to clinical implementation.

The economic and accessibility divide is another important biotechnology concern. The cost of developing genetic modification technologies is high, particularly in the fields of agriculture and medicine. Because of this, wealthier countries and private companies are at the forefront of many advancements in synthetic biology, gene treatments, and genetically modified organisms, which may limit access for developing nations. For example, the high cost of seeds and intellectual property restrictions imposed by biotech companies can often hinder the adoption of genetically modified crops in less developed regions, even though they have the potential to alleviate food poverty in many parts of the world. Closing this gap will require international cooperation, more affordable technology, and systems to ensure that biotechnological innovations are accessible to all, particularly those in developing countries.

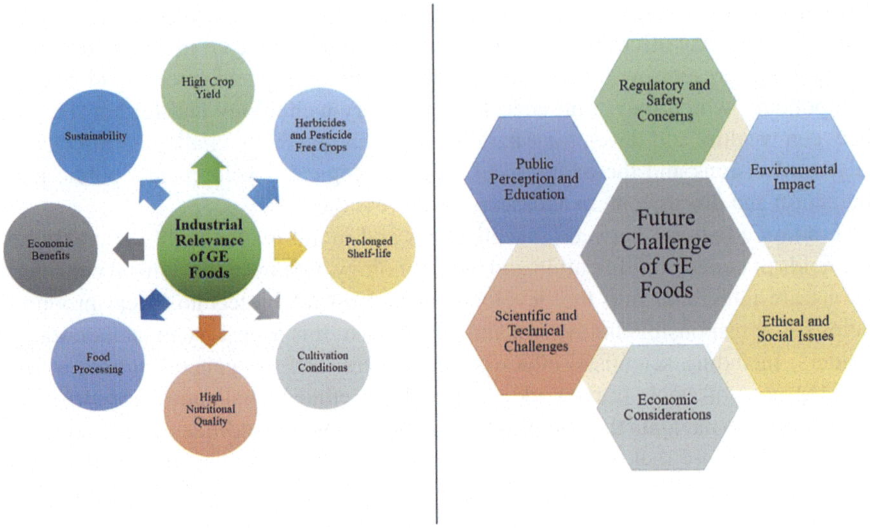

Fig. 6.5 Challenges and prospects of genetically modified (GM) foods. (Yadav et al., 2024)

Future prospects for genetic alteration and biotechnology are very promising. The creation of precision gene-editing tools, including prime editing and base editing, allows for even more precise modification of genetic material, reducing the likelihood of unforeseen repercussions and increasing the range of diseases that may be treated. Synthetic biology is expected to play a key role in the growth of bio-based economies, as modified organisms could reduce our reliance on fossil fuels and provide more sustainable ways to produce food, energy, and materials. Furthermore, biocomputing and bio-manufacturing utilizing modified cells and enzymes have the potential to transform industrial processes and produce cleaner, more efficient production systems. The bio-electrochemical conversion of waste into fuel, in conjunction with advancements in enzyme engineering, may provide innovative methods for waste management and the production of renewable energy.. However, it will be essential to ensure ethical use, equitable access, and stringent regulatory oversight as the topic advances and affects society (Fig. 6.5).

7 Conclusion

Some of humanity's most pressing issues, including food security, health care, and environmental sustainability, could be resolved by the innovative fields of biotechnology and genetic modification. Researchers are using cutting-edge techniques like CRISPR-Cas9 and synthetic biology to find new solutions. It might lead to more resilient crops, effective cures for inherited diseases, and ecologically friendly industrial operations. The diverse applications of genetic modification, ranging from bioremediation and medicinal breakthroughs to agricultural improvements, illustrates the versatility and impact of these technologies across numerous domains. However, the rapid advancement of genetic modification and biotechnology presents a variety of moral and legal challenges that must be appropriately addressed. Ensuring the safety of genetically modified organisms and biotechnological products, as well as fostering public trust in these technologies, are essential. Scientists, policymakers, and the general public must constantly debate the ethical implications of gene editing, particularly in relation to equitable access, intellectual property rights, and human germline modification.. It will be essential to address these concerns through transparent regulatory frameworks and inclusive discussions in order to ensure that the benefits of biotechnology are realized without compromising safety or ethical standards.

The future of genetic modification and biotechnology appears bright. As these technologies continue to develop, guiding their development and application will need fostering an ethical and responsible mindset. If the drive for scientific advancement is matched with ethical considerations and public involvement, society will be able to fully exploit genetic modification and biotechnology. This will pave the way for a more equitable and sustainable future. By addressing challenges and taking advantage of possibilities, we may use these powerful tools to enhance agriculture, health, and the environment for coming generations.

References

Cockburn, I. M., & Stern, S. (2010). Finding the endless frontier: lessons from the life sciences innovation system for technology policy. *Capitalism and Society, 5*(1).

Kumar, K., Gambhir, G., Dass, A., Tripathi, A. K., Singh, A., Jha, A. K., Yadava, P., Choudhary, M., & Rakshit, S. (2020). Genetically modified crops: current status and future prospects. *Planta, 251*(4), 91.

Langer, R., & Sharma, S. (2020). The blessing and curse of biotechnology: A primer on biosafety and biosecurity, carnegie endowment for international peace. https://www.google.com/search?q=https%3A%2F%2Fcarnegieendowment.+org+%E2%80%A6.&rlz=1C1GCEU_enIN1080IN1080&oq=https%3A%2F%2Fcarnegieendowment.+org+%E2%80%A6.&gs_lcrp=EgZjaHJvbWUyBggAEEUYOTIICAEQABgWGB4yCggCEAAYChgWGB4yCAgDEAAYFhgeMggIBBAAGBYYHjIHCAUQABjvBTIHCAYQABjvBdIBBzQ3M2owajmoAgCwAgE&sourceid=chrome&ie=UTF-8

Yadav, V., Pal, D., & Poonia, A. K. (2024). A study on genetically engineered foods: Need, benefits, risk, and current knowledge. *Cell Biochemistry and Biophysics, 82*(3), 1931–1946.

Part III
Integration of Bioprocessing and Traditional Processing

Chapter 7
Benefits and Challenges of Integration in Food and Bioprocessing

Abdul Sattar Jatoi, Arshad Iqbal, and Bazil Owais Nasir

1 Introduction

The use of biotechnology in food and bioprocessing represents a significant advancement in the quest for efficient and sustainable food production systems. As the world's population grows, so does the need for food, placing additional pressure on agricultural systems. Traditional farming methods face a number of challenges, including resource scarcity, climate change, and the need for better food quality and safety. Since the food industry can provide innovative solutions that boost productivity, reduce waste, and create healthier food options by applying biotechnology, integration is an essential area of study for future food security.

The vast field of biotechnology encompasses a wide range of techniques and processes that can be applied to enhance food production, including genetic engineering, fermentation, and enzyme technology. By integrating these biotechnology techniques with traditional food manufacturing, manufacturers may optimize numerous stages of the food supply chain, from obtaining raw materials to distributing final products. Fermentation techniques, for instance, can be used to create foods like cheese and yogurt, improve the nutritional value of crops, and create functional foods that offer health benefits beyond basic subsistence. This type of integration fosters innovation and interdisciplinary collaboration, enabling new food items that meet evolving consumer demands.

There are major environmental benefits to using biotechnology in food and bioprocessing. By improving the efficiency of resource usage, including raw materials,

A. S. Jatoi (✉) · A. Iqbal · B. O. Nasir
Department of Chemical Engineering, Dawood University of Engineering and Technology, Karachi, Pakistan
e-mail: abdul.sattar@duet.edu.pk

A. S. Jatoi (ed.), *Integrated Food and Bioprocessing*,
https://doi.org/10.1007/978-3-031-87209-9_7

energy, and water, biotechnology can reduce the ecological impact of food production. Additionally, bioprocessing techniques can assist a circular economy by converting food and agricultural waste into beneficial products like compost or biofuels. By cutting waste and creating opportunities for sustainable practices in the food industry, this approach promotes resilience in the face of environmental challenges.

Despite the numerous advantages, there are some challenges associated with the application of biotechnology in food and bioprocessing. Technical challenges, intricate rules, and issues with public acceptance could all impede progress. Integrating different biotechnology approaches into existing food systems requires careful planning, expertise, and financial resources. Additionally, manufacturers may find it challenging to ensure food safety and transparency while navigating the regulatory landscape surrounding genetically modified organisms (GMOs). A number of problems need to be fixed before biotechnology advancements may be effectively implemented in the food sector.

The application of biotechnology in food and bioprocessing has enormous promise for enhancing global food security and transforming food systems in the future. Participants in the food industry can work together to create a food supply chain that is more robust, effective, and sustainable by utilizing innovative techniques and state-of-the-art technologies. Policymakers, researchers, and business leaders must collaborate to develop frameworks that enable the appropriate use of biotechnology in order to address ethical concerns and preserve public trust. Ultimately, successful integration will require a holistic approach that balances environmental preservation, social responsibility, and technological advancement, leading to a healthier and more sustainable future for everybody (Fig. 7.1).

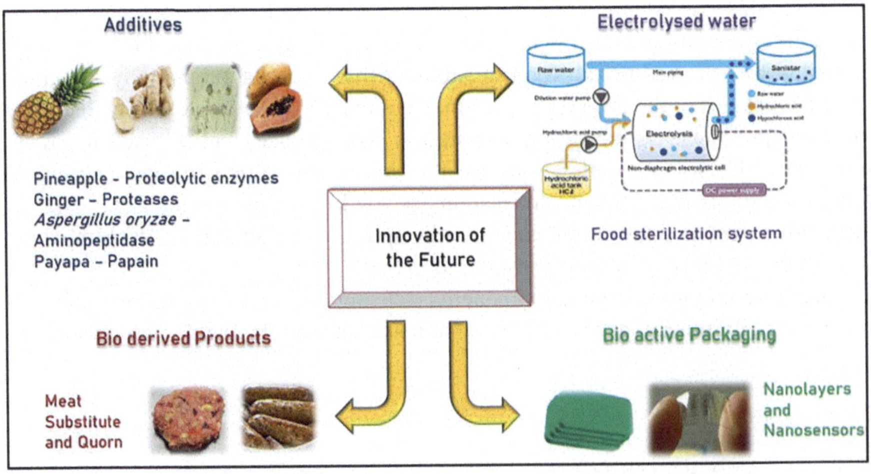

Fig. 7.1 Food biotechnology: innovations and challenges. (Cabrera-Barjas et al., 2022)

2 Benefits of Integration in Food and Bioprocessing

Using biotechnology in food and bioprocessing has a number of benefits that increase output, sustainability, and product quality. From crop growth to the delivery of completed food items to consumers, these benefits span the whole food production and processing lifecycle. Some of the primary advantages of this integration are as follows:

1. Improved Efficiency

Incorporating biotechnology technologies into food production significantly boosts efficiency by streamlining processes. Techniques like fermentation and enzymatic processing can reduce energy and resource use by improving the extraction and production of food ingredients. For example, enzyme technology can boost the yields of sugars from starches or proteins from plant sources, leading to more effective processing methods. Furthermore, waste may be decreased at every level of the production process using biotechnology and automation, which enables producers to boost output while cutting costs.

2. Enhanced Food Quality and Safety

Biotechnology plays a major role in improving food products' safety and quality. Through genetic modification and selective breeding, crops can be altered to have improved nutritional profiles, longer shelf lives, and superior flavors. Biofortified foods, such as Golden Rice, provide essential vitamins and minerals to compensate for nutritional deficiencies. Additionally, bioprocessing techniques can reduce harmful microorganisms and hazardous compounds in food products. For example, fermentation creates conditions that inhibit the growth of spoilage organisms while also enhancing flavor, making food safer and better.

3. Sustainability and Environmental Benefits

Environmental sustainability is promoted by the application of biotechnology in food and bioprocessing, which reduces the ecological footprint of food production. Biotechnological developments lessen the adverse effects of natural resources on ecosystems by facilitating their more efficient usage, such as water and land. For example, genetically modified crops that require fewer inputs, such as fertilizer and pesticides, may lead to more sustainable farming practices. Furthermore, food waste can be converted into beneficial byproducts like animal feed or biofuels through bioprocessing, which will lessen landfill waste and promote a circular economy.

4. Innovation and Product Development

Integrating biotechnology promotes innovation in the development of food products, allowing manufacturers to create new foods that cater to changing consumer preferences. Novel foods and flavors can be created via techniques like synthetic biology, even though fermentation can provide unique textures and flavors. The development of functional foods—such as probiotics that promote gut health—that provide health benefits beyond basic sustenance is another example of this innovation. By staying ahead of consumer expectations and market trends, businesses may differentiate themselves and take advantage of new market opportunities.

5. Economic Benefits

Economic benefits can be obtained by both farmers and consumers through the incorporation of biotechnology into food and bioprocessing. By improving efficiency and reducing manufacturing costs, businesses can increase profitability and potentially lower consumer prices. Furthermore, the application of biotechnological techniques may lead to new markets and job opportunities in the food and biotechnology sectors. For instance, advancements in plant-based proteins and other food sources can promote economic growth and meet the rising need for sustainable food options.

In conclusion, applying biotechnology to food and bioprocessing has a number of benefits that enhance quality, sustainability, innovation, and productivity. By taking use of these advantages, food industry stakeholders may develop more resilient and effective food systems that meet the needs of a changing environment and a growing global population (Fig. 7.2).

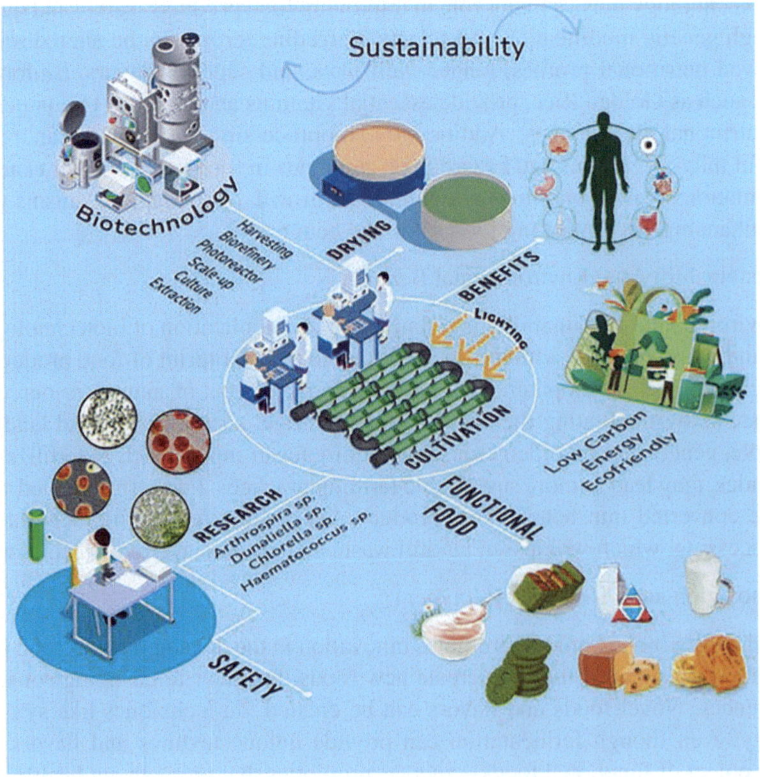

Fig. 7.2 Challenges in functional food products. (Çelekli et al., 2024)

3 Challenges of Integration in Food and Bioprocessing

While incorporating biotechnology into food and bioprocessing has many benefits, there are also certain problems that must be fixed for the project to be carried out properly. These challenges, which might derive from technical, regulatory, economic, ethical, and social factors, may affect the overall effectiveness of integrated systems. Some significant challenges are as follows:

1. Technical Challenges

Integrating different biotechnology technologies into existing food systems might be technically difficult. Variability in raw materials affects bioprocess efficiency, requiring precise control over conditions to ensure consistent output quality. Compatibility problems may also occur since not all bioprocessing methods can be fully integrated. The need for qualified personnel who understand both food science and biotechnological processes adds even more complexity, as training and experience are critical for successful implementation.

2. Regulatory Hurdles

The legal framework governing biotechnology and genetically modified organisms (GMOs) can be complicated and varies substantially between countries and geographical areas. Navigating these limitations can be a significant challenge for companies looking to integrate biotechnology into their operations. Regulatory permits often need thorough safety assessments, which can be costly and time-consuming. Furthermore, disparate regulatory standards might complicate the integration process for global supply chains by causing misunderstandings and impeding the international trade of biotechnology products.

3. Economic Barriers

Because new biotechnological processes require a large initial investment, many producers—especially small and medium-sized businesses—find it difficult to integrate them. The high costs of research and development, facility upgrades, and technology acquisition can deter businesses from adopting new techniques. Furthermore, the economic viability of integrated bioprocessing methods is also influenced by customer demand and market acceptance, which can alter and impact return on investment.

4. Consumer Acceptance and Ethical Considerations

There are still several barriers to the public's acceptance and perception of biotechnological developments in food production. Many consumers have doubts about genetically modified organisms (GMOs) and bioprocessed foods due to concerns about the impact on their safety, health, and the environment. Misinformation and ignorance of biotechnology might exacerbate these concerns and lead to resistance to integrated approaches. Ethical concerns like patenting and genetic resource ownership must be addressed in order to foster public confidence and ensure the responsible use of biotechnology in food systems.

5. Skills and Knowledge Gaps

The successful integration of biotechnology in food and bioprocessing requires a workforce with the necessary skills and expertise. However, a lack of education and experience may make it more difficult to embrace new strategies. Closing this gap is essential to ensure that employees are informed on both traditional food production methods and modern biotechnology approaches. This may mean developing specialized training programs and promoting collaboration between academic institutions and industry participants in order to produce a skilled workforce.

6. Case Studies and Examples

The integration of biotechnology into food and bioprocessing has been illustrated by a number of successful case studies, which also highlight innovative approaches and the benefits of these endeavors. Here are few examples that stand out:

1. **Impossible Foods: Plant-Based Meat Alternatives**

Impossible Foods, a company that specializes in creating plant-based meat alternatives with the taste and consistency of animal flesh, is among the best instances of biotechnology integration in the food business. The company's flagship product, the Impossible Burger, uses soy leghemoglobin, a genetically modified protein that mimics the heme in animal blood, to give it a meat-like flavor. By combining biotechnology with food processing techniques, Impossible Foods has successfully developed a product that not only addresses the ethical and environmental concerns related to the production of meat but also appeals to a growing consumer base searching for sustainable food alternatives. By using state-of-the-art bioprocessing technology, the company's product has a reduced carbon footprint than beef produced the traditional way.

2. **Bioconversion of Food Waste: The Upcycling Movement**

As a way to combat food waste and create useful products, upcycling has grown in popularity. Regrained and Wasted are two companies that use biotechnology and bioprocessing to create new, sustainable food items from food industry waste. Regrained, for instance, transforms waste grain from the brewing industry into nutritious flour that can be used in a range of baked goods. This partnership not only reduces trash but also offers consumers healthier options and a new revenue stream for breweries. These bioconversion processes are excellent illustrations of how biotechnology may be used in food systems to encourage innovation and sustainability.

3. **Enzymatic Processing in Dairy Industry: DSM's Maxiren**

With its product Maxiren®, an enzyme that enhances cheese production, the global science-based company DSM has successfully integrated enzymatic bioprocessing into the dairy industry. DSM uses genetically modified microorganisms to produce this enzyme, which reduces processing time and improves cheese yield and quality. Biotechnology can help dairy companies boost efficiency, decrease waste, and streamline their operations. This illustration demonstrates how biotechnology can enhance traditional food processing methods, boosting the sustainability and output of the dairy sector.

4. **Genetically Modified Crops: Bt Cotton and Bt Corn**

One example of the benefits of integrating biotechnology into agriculture is the use of genetically modified crops, such as Bt corn and Bt cotton. Bt crops are engineered to express a protein from the Bacillus thuringiensis (Bt) bacterium, which gives resistance against particular pests, hence reducing the need for chemical pesticides. The implementation of this biotechnology approach has increased agricultural yields and reduced environmental impact because farmers can now produce more with fewer chemical inputs. Case studies in countries like the US and India show that the adoption of Bt crops has significantly raised farmers' incomes, demonstrating the positive economic consequences of integrating biotechnology into agriculture and food production.

5. **Fermentation Technology: Beyond Meat's Use of Pea Protein**

Beyond Meat is another company that is leading the way in plant-based meat alternatives by fusing fermentation and protein processing technologies. By employing pea protein as its primary ingredient, Beyond Meat has produced a product that nearly mimics the flavor and texture of beef. Customers searching for meat substitutes will find the texture and flavors the company creates thanks to its innovative use of fermentation techniques. This integration not only meets the growing demand for plant-based substitutes, but it also shows how biotechnology can be applied to create sustainable food products that meet a variety of dietary needs.

4 Future Prospects

The use of biotechnology in food and bioprocessing is anticipated to have a significant impact on the food industry's future. As global problems like resource scarcity, climate change, and population growth worsen, innovative and sustainable food production methods are becoming more and more significant. Some noteworthy prospects for the future of biotechnology integration in food and bioprocessing include the following:

1. **Advancements in Synthetic Biology**

Synthetic biology is an emerging field that combines biology and engineering to produce new biological systems, tools, and components. This innovative method has the potential to revolutionize the food manufacturing business by enabling the development of novel flavors and ingredients that meet evolving consumer preferences. For example, synthetic biology can assist in producing protein substitutes that taste and feel like animal goods, providing sustainable alternatives to traditional livestock farming. As technology develops, this integration should lead to more customized and health-conscious food offerings.

2. **Precision Agriculture and Smart Farming**

Combining biotechnology with precision agricultural and smart farming technologies could significantly improve food production's sustainability and efficiency. Using biotechnology developments, artificial intelligence, and data analytics, farmers may monitor crop health, control resource use, and tailor interventions to specific needs. For example, biotechnology can improve crop resilience to environmental stresses, while precision agriculture technologies can reduce inputs like water and fertilizer. A more sustainable food system will result from this all-encompassing strategy, which will increase yields while reducing the negative environmental effects of farming methods.

3. **Biotechnology for Food Safety and Quality**

Future advancements in biotechnology will enhance food safety and quality by better identifying and preventing foodborne toxins and infections. It will be feasible to produce crops and food products that are inherently resistant to contamination and deteriorating thanks to developments in genetic engineering and microbiome research. For instance, due of their improved nutritional profiles and longer shelf lives, genetically modified fruits and vegetables can increase food availability and reduce food waste. In addition to ensuring safer food supply chains, rapid pathogen testing techniques will meet consumer demands for high-end products.

4. **Circular Economy and Waste Reduction**

As the world transitions to a more sustainable future, the application of biotechnology in food and bioprocessing will be crucial to promoting a circular economy. By developing processes that convert food waste into valuable byproducts like biofuels, animal feed, or bioplastics, businesses may cut waste and increase resource efficiency. In keeping with global sustainability goals, this approach addresses the pressing issue of food waste, which is estimated to be a major contributor to greenhouse gas emissions globally. Biotechnological approaches to waste management will lead to a more sustainable and effective food system.

5. **Regulatory and Consumer Engagement**

Future food and bioprocessing biotechnology integration will also depend on effective regulatory frameworks and consumer participation. As biotechnology advances, regulatory bodies will need to adapt their approaches to preserve safety and foster innovation. It will take open communication and educational initiatives to increase public acceptance and trust in biotechnology advances. Engaging consumers in discussions regarding the benefits and drawbacks of biotechnology will empower them to make informed choices and encourage the expansion of moral business conduct (Fig. 7.3).

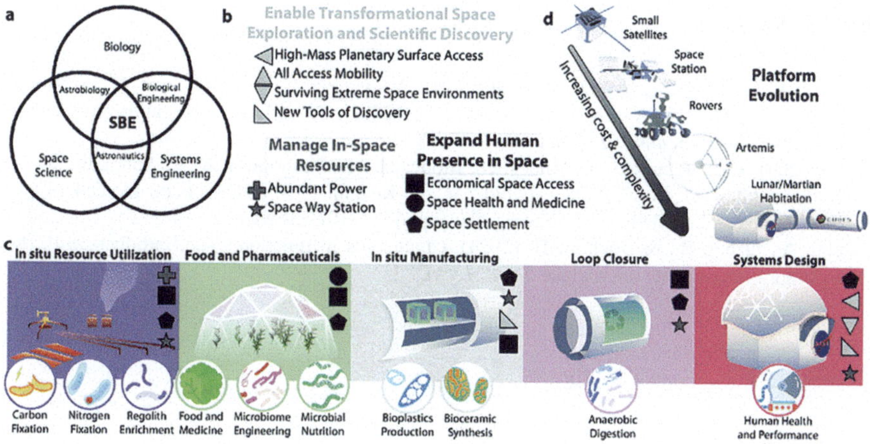

Fig. 7.3 Space bioprocess engineering on the horizon. (Berliner et al., 2022)

5 Conclusion

The use of biotechnology in food and bioprocessing has fundamentally altered our understanding of food production and sustainability. By using advanced biotechnology techniques, the food industry can address significant problems like as resource scarcity, environmental sustainability, and the growing demand for healthy, high-quality food. A variety of sustainable food options that meet shifting consumer wants are made possible by this integration, which also increases the efficiency of food production processes and encourages creativity in product creation. But there are obstacles standing in the way of this integration's full benefits. It is crucial to carefully manage technical complexity, regulatory obstacles, economic constraints, and consumer acceptance concerns in order to ensure the proper implementation of biotechnological innovations. To address these concerns and foster an environment that promotes innovation while maintaining safety and moral standards, industry players, lawmakers, academics, and consumers must collaborate. Future prospects for food biotechnology and bioprocessing appear promising. As precision agriculture, synthetic biology, and waste reduction technologies continue to progress, the possibility for creating a more robust, effective, and sustainable food system will increase. By embracing the integration of biotechnology and cooperating to address its challenges, we can build a more sustainable and healthy future for all. This will ensure that future food systems are ready to meet the needs of a growing global population while protecting the environment.

References

Berliner, A. J., Lipsky, I., Ho, D., Hilzinger, J. M., Vengerova, G., Makrygiorgos, G., McNulty, M. J., Yates, K., Averesch, N. J., & Cockell, C. S. (2022). Space bioprocess engineering on the horizon. *Communications Engineering, 1*(1), 13.

Cabrera-Barjas, G., Banerjee, A., Valdes, O., Moncada, M., Sirajunnisa, A. R., Surendhiran, D., Ramakrishnan, G., Rani, N. S., Hamidi, M., & Kozani, P. S. (2022). Food biotechnology: Innovations and challenges. In *Future foods* (pp. 697–719). Elsevier.

Çelekli, A., Özbal, B., & Bozkurt, H. (2024). Challenges in functional food products with the incorporation of some microalgae. *Food, 13*(5), 725.

Chapter 8
Case Studies in Integrated Bioprocesses

Abdul Sattar Jatoi, Arshad Iqbal, Bazil Owais Nasir, and Muhammad Azam Usto

1 Introduction

The use of biotechnology in food and bioprocessing is transforming food production by offering innovative solutions to some of the most significant issues confronting the industry. With the world's population growing and the need for sustainable food supply increasing, there has never been a more urgent need for efficient and effective food production methods. The efficiency, sustainability, and quality of food systems can be enhanced by integrated procedures, which combine conventional techniques with various biotechnology approaches. Several case studies that show the successful implementation of integrated processes in the food and bioprocessing sectors are examined in this chapter. Case studies can be used to comprehend the practical impacts of integrating biotechnology into agricultural production. They provide real-world examples that show how innovative approaches can be applied to address issues including resource scarcity, environmental impact, and changing consumer preferences. By examining specific instances of effective integration, stakeholders can gain a better understanding of the benefits, drawbacks, and overall impacts of these tactics. From waste reduction strategies to plant-based alternatives, this chapter aims to present a broad range of case studies that illustrate the versatility of biotechnology in numerous contexts. Each case study in this chapter was selected based on a set of criteria, including sustainability, innovation, and economic benefit. These studies reflect several industries within the food industry and demonstrate the variety of biotechnology uses, from creating genetically modified crops to repurposing food waste. By providing a comprehensive overview of integrated processes that have worked, the chapter aims to demonstrate how

A. S. Jatoi (✉) · A. Iqbal · B. O. Nasir · M. A. Usto
Department of Chemical Engineering, Dawood University of Engineering and Technology, Karachi, Pakistan
e-mail: abdul.sattar@duet.edu.pk

A. S. Jatoi (ed.), *Integrated Food and Bioprocessing*,
https://doi.org/10.1007/978-3-031-87209-9_8

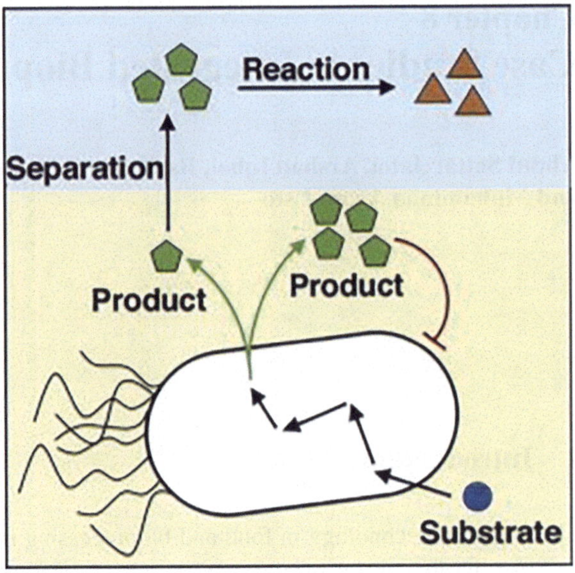

Fig. 8.1 Aiding and expanding microbial biofuel/biochemical production. (Flores et al., 2019)

biotechnology has the potential to drastically change food production practices. This analysis of case studies will reveal recurrent patterns and lessons with wider relevance in addition to highlighting particular achievements. As we look at the particular case studies, it is important to recognize the transformative potential of integrated processes in affecting the future of food systems. The corporate and innovation stories in this chapter will demonstrate how biotechnology and traditional methods of food production can be combined to provide more resilient, efficient, and sustainable food systems. By implementing these integrated strategies, the food industry may be better equipped to handle the challenges posed by a changing environment, consumer demands, and the need for responsible resource management (Fig. 8.1).

2 Case Study Selection Criteria

The selection of case studies is an essential first step in understanding the successful integration of biotechnology into food and bioprocessing. To ensure a comprehensive and meaningful analysis of integrated processes, certain criteria have been established to guide the selection of relevant case studies. These criteria, which address topics like innovation, sustainability, economic impact, application, and diversity, enable a comprehensive representation of the region. The following is a list of the primary selection criteria for this chapter:

2.1 Innovation

The first criterion focuses primarily on the degree of innovation demonstrated in the case studies. This includes the application of novel biotechnology products, processes, or strategies that significantly enhance traditional food production methods. The most recent developments in the field are exemplified by case studies that highlight innovative methods, such as the use of fermentation technology, enzymatic processing, or synthetic biology. By emphasizing innovation, the food and bioprocessing industries may better understand how biotechnology can drive progress and provide competitive advantages.

2.2 Sustainability

Sustainability is a key component of modern food production, and selected case studies must demonstrate a strong commitment to environmental responsibility and resource efficiency. This criterion evaluates how integrated activities contribute to waste reduction, energy conservation, and a reduction in the ecological impact of food production. Case studies demonstrating efficient waste management strategies, such recycling leftovers or using the ideas of the circular economy, demonstrate how biotechnology can create more sustainable food systems.

2.3 Economic Impact

The economic effects of integrating biotechnology into food and bioprocessing must be taken into account in order to assess the overall feasibility and success of these advancements. Well-chosen case studies should illustrate quantifiable economic benefits, such as cost savings, increased profitability, or enhanced market competitiveness. This criterion helps demonstrate that producers and other food industry stakeholders get both financial and environmental advantages from the use of integrated processes.

2.4 Applicability and Scalability

To ensure that the case studies are relevant to a broad audience, they must exhibit application and scalability in a range of contexts. This criterion assesses how well the integrated processes are implemented or adjusted for different market niches, geographical areas, or food production systems. By looking at case studies that demonstrate flexible and replicable solutions, other industry stakeholders can learn from successful instances and consider how similar innovations can be applied in their own operations.

2.5 Diversity of Approaches

A broad range of case studies is necessary to completely convey the intricate nature of integrated systems in food and bioprocessing. This criterion encourages the inclusion of case studies from a variety of businesses, including waste reduction, enzymatic processing, genetically modified crops, and plant-based alternatives. Through the presentation of diverse biotechnology techniques and applications, the chapter provides a comprehensive overview of the field and highlights the potential and flexibility of integrated processes throughout the food sector.

3 Case Study 1: Impossible Foods

In 2011, Patrick Brown founded the biotechnology company Impossible Foods with the intention of creating sustainable alternatives to animal-based foods. In order to reduce the environmental impact of food production, the company plans to provide delicious plant-based meat substitutes that will appeal to both vegetarians and meat eaters. By focusing on innovation and sustainability, Impossible Foods aims to address pressing problems associated with traditional animal agriculture, such as greenhouse gas emissions, water use, and land degradation.

Integrated Bioprocesses Used to Create Plant-Based Meat Alternatives
Impossible Foods' product development is centered on integrated bioprocesses and innovative biotechnology. The company's hallmark meal, the Impossible Burger, is primarily made of plant-based ingredients like pea protein, coconut oil, and potato starch. An important innovation is the incorporation of soy leghemoglobin, a protein derived from the roots of soy plants that closely resembles heme, the iron-containing molecule that gives meat its unique flavor and color. Through genetic engineering, Impossible Foods has successfully produced soy leghemoglobin in yeast, allowing for its mass production without the need for animal sources.

The production process consists of several processes, such as fermentation, extrusion, and formulation. During the fermentation phase, yeast is cultivated in bioreactors to produce soy leghemoglobin. The extruder then processes the combination, giving it a meat-like texture that mimics the mouthfeel and juiciness of beef. This combined method allows for efficient scalability and production while maintaining the final product's flavor and quality. Impossible Foods has created an ecologically benign and consumer-friendly meat alternative by utilizing biotechnology and advanced manufacturing techniques (Fig. 8.2).

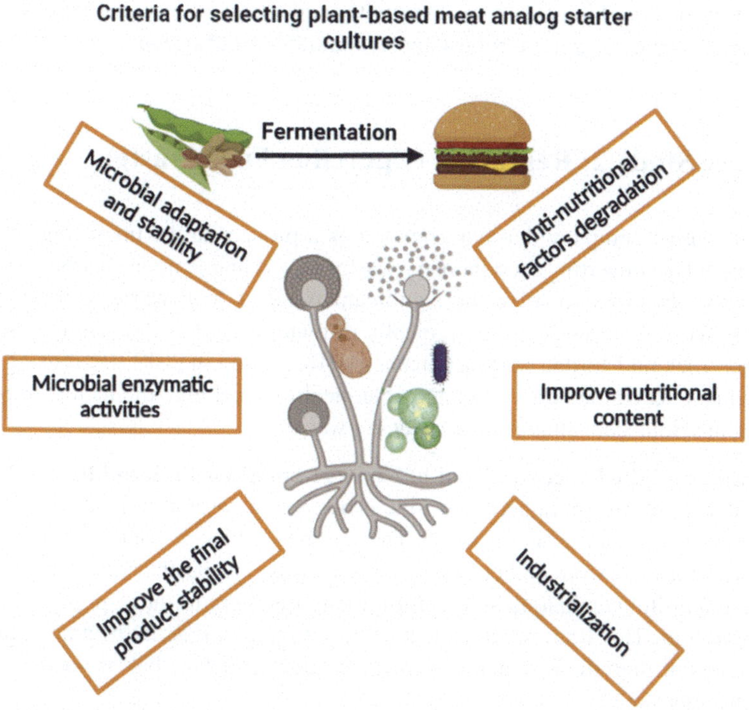

Fig. 8.2 Significance of fermentation in plant-based meat analogs. (Elhalis et al., 2023)

Impact on Sustainability, Consumer Preferences, and the Food Industry

The impact of Impossible Foods on sustainability is significant. The Impossible Burger requires about 95% less land, 74% less water, and emits 87% fewer greenhouse gases than conventional beef production. This reduction in resource usage demonstrates how plant-based alternatives could help mitigate the environmental issues caused by animal husbandry.

According to consumer preferences, Impossible Foods has successfully attracted a wide spectrum of clients, including both flexitarians and die-hard meat eaters. The company has collaborated with big-box stores and restaurants to make its products widely accessible, which has contributed to the rise in popularity of plant-based diets. By becoming a market leader in the alternative protein area, Impossible Foods has changed consumer behavior and inspired other companies to take similar actions.

The introduction of the Impossible Burger has also sparked a broader conversation about the future of food production and the application of biotechnology to global challenges. Impossible Foods' success demonstrates how integrated bioprocesses have the potential to drastically change the food industry as consumers become more aware of how their dietary choices affect both their health and the environment. When everything is said and done, Impossible Foods is a compelling illustration of how biotechnology, creativity, and sustainability can be effectively integrated to produce a more robust and accountable food system.

4 Case Study 2: Regrained (Upcycling Food Waste)

One innovative company that has emerged as a pioneer in the upcycling sector is Regrained. By converting food industry wastes into valuable items, it aims to reduce food waste. In order to solve the significant issue of food waste in the brewing industry, where excess grains are typically thrown away after the brewing process, Dan Kurzrock and Jordan Schwartz launched Regrained in 2013. By reusing these by-products, Regrained creates sustainable, healthy food ingredients that appeal to health-conscious consumers while reducing waste.

Description of the Processes Used to Convert Spent Grain into Flour
The initial stage in Regrained's creative process is gathering leftover grains from local breweries after the beer is made. These grains, which are rich in protein, fiber, and other nutrients, would otherwise be thrown away. The business uses an advanced drying and milling technique to transform the leftover grains into Upcycled Flour, a premium flour. This flour retains much of the original grains' nutritional value and is an essential ingredient in many culinary products, such as baked goods, protein bars, and snacks.

The upcycling process consists of two steps: first, the leftover grains are collected, and then they are dried to reduce moisture content and prevent rotting. The dried grains are then crushed into flour to ensure that the final product meets food safety standards. The resulting Upcycled Flour can be marketed as a nutritious alternative to traditional flours, drawing in clients who respect wellbeing and sustainability. By converting a by-product into a versatile substance, Regrained demonstrates how bioprocessing may transform waste into valuable products.

Environmental and Economic Benefits of Upcycling Food Waste
Regrained's upcycling of leftover grains has significant environmental and economic implications. By preventing leftover grains from ending up in landfills, the company actively contributes to the reduction of food waste, which is a major source of greenhouse gas emissions. It is estimated that 8–10% of global greenhouse gas emissions are caused by food waste. By converting this garbage into a useful product, Regrained helps breweries reduce their waste disposal costs while also lessening their environmental effect.

Economically speaking, Regrained's concept strengthens local economies and provides breweries with new revenue streams. By partnering with breweries to source leftover grains, Regrained fosters relationships that benefit both parties. Breweries can reduce waste and establish a reputation for their eco-friendly practices, while Regrained can use the abundant supply of leftover grains to make its Upcycled Flour. The economic feasibility of upcycling food waste is further demonstrated by Regrained, which enters a lucrative market by offering a wholesome product that meets the rising demand for sustainable and health-conscious food options.

5 Case Study 3: DSM's Maxiren® in Dairy Production

Royal DSM is a global science-based company that specializes in health, nutrition, and materials. Due to its commitment to sustainability and innovation, DSM has developed several solutions to enhance food production and product quality. In the dairy industry, where demand for high-quality cheese and yogurt is rising, production process optimization is essential to meeting consumer needs while minimizing environmental impact. DSM's Maxiren® product line, which comprises innovative coagulants and enzymes, aims to address the challenges faced by dairy producers while enhancing the quality and productivity of cheesemaking.

Description of the Use of Enzymatic Bioprocessing in Cheese Production
The coagulation and curd formation processes required to make cheese are significantly improved by a line of enzyme preparations known as Maxiren®. Traditional cheesemaking uses rennet, an enzyme derived from animals, to coagulate milk. In response to the growing demand for vegetarian-friendly options, Maxiren® offers a microbiological substitute that expands the adaptability of cheese production. The Maxiren® enzymes are produced by fermentation, which ensures a consistent supply and reliable operation.

By employing Maxiren® enzymes in the dairy production process, cheese makers may produce products with the necessary textures, flavors, and functionalities. These enzymes shorten processing times and increase curd production by accelerating coagulation. Additionally, Maxiren® makes it easier to regulate the fermentation process, which enhances the flavor profiles of many cheese varieties. This integrated enzymatic bioprocess contributes to the production of high-quality dairy products that meet consumer demands while also increasing operational effectiveness.

Outcomes Related to Efficiency, Quality Improvement, and Market Competitiveness

The use of Maxiren® in cheesemaking has yielded noteworthy outcomes for dairy producers. By making the coagulation process simpler, dairy companies may increase production capacity and reduce energy use. This improved efficiency lowers manufacturing costs and increases profitability by enabling businesses to better match consumer needs.

Additionally, producers are now more competitive in the dairy sector thanks to their ability to regularly produce high-quality cheese. By using Maxiren®, producers may make cheeses that meet both regulatory requirements for vegetarian-friendly options and consumer preferences for artisanal, premium cheese products. This emphasis on quality and sustainability gives dairy producers a competitive edge in addition to satisfying consumer demand.

In conclusion, DSM's Maxiren® is an excellent illustration of how enzymatic bioprocessing has the potential to completely transform the dairy industry. Maxiren® offers innovative solutions that increase productivity and quality while solving significant challenges faced by cheese producers and supporting the shift toward sustainable food production. Meeting consumer expectations and advancing the expansion of the dairy business as it undergoes further transformation will need the adoption of integrated bioprocesses, including those offered by DSM.

6 Case Study 4: Genetically Modified Crops (Bt Cotton and Bt Corn)

Genetically modified organisms (GMOs) have revolutionized agriculture by enabling the development of crops with improved nutritional value, resistance to pests, and tolerance to herbicides. One of the most common applications of genetic modification in agriculture is the introduction of **Bacillus thuringiensis** (Bt) genes into corn and cotton plants. Certain insect pests are poisoned by proteins produced by a soil bacteria known as Bt. By incorporating these genes into crops like Bt corn and cotton, farmers can significantly reduce their reliance on chemical pesticides, boosting agricultural yields and reducing their adverse environmental consequences.

Examination of the Impact of Bt Crops on Pest Resistance and Crop Yields

In many countries, Bt corn and cotton have been eagerly adopted because to their success in managing pest populations. By granting resistance to significant pests like the cotton bollworm, the Bt gene is incorporated into Bt cotton, reducing the requirement for pesticide applications. This insect resistance has led to a notable improvement in cotton production.; Research has indicated a 20–30% increase in productivity compared to traditional cotton genotypes. Similarly, it has been shown that Bt corn increases output and reduces losses by effectively combatting pests like the European corn borer.

Bt crops promote more ecologically friendly farming practices in addition to raising yield. By reducing the need for chemical pesticides, Bt crops help lessen pollution in the environment and harm to beneficial insects and other non-target species. Farmers also earn financially from the adoption of Bt crops since increased yields and reduced pesticide costs result in increased profitability. All things considered, the creation of Bt cotton and corn represents a significant advancement in agricultural biotechnology and provides workable solutions to insect control issues.

Analysis of Socio-Economic Benefits for Farmers and Environmental Considerations

Significant socioeconomic benefits are offered by Bt cotton and corn, particularly to smallholder farmers in poor countries. These crops have allowed farmers to raise their standard of living by increasing their income and reducing crop losses due to pests. Studies show that many farmers' net incomes have improved by 50% as a result of the adoption of Bt cotton, enabling them to invest in healthcare, education, and other aspects of their homes.

However, the use of Bt crops is controversial and raises important environmental and socioeconomic concerns. One problem is the potential for insect resistance to develop, which would reduce the long-term effectiveness of Bt crops. Two essential integrated pest management (IPM) strategies that assist lessen this risk and ensure the long-term sustainability of Bt technology are crop rotation and the use of non-Bt refuges. Furthermore, the economic dependence on a limited number of genetically modified crops may jeopardize biodiversity and conventional farming practices.

In conclusion, Bt corn and Bt cotton are excellent illustrations of how agriculture might be transformed by genetically modified crops. By successfully fending off pests and boosting yields, these crops have enhanced farmer livelihoods and encouraged more environmentally friendly farming practices. As the agricultural landscape evolves, careful GMO management and ongoing study will be necessary to optimize the advantages of GMOs while addressing the associated environmental and social problems. In order to develop a robust and sustainable food supply, the Bt crop scenario highlights how important it is to find a balance between biotechnology innovation and responsible stewardship.

7 Case Study 5: Beyond Meat's Use of Fermentation Technology

Introduction to Beyond Meat and Its Product Offerings

In 2009, Ethan Brown established Beyond Meat, a well-known company in the plant-based protein sector, with the goal of creating delicious and sustainable alternatives to traditional animal-based meat products. The company's primary objective is to address the health, environmental, and animal welfare issues associated with meat consumption. Beyond Beef has developed a range of innovative products, including ground beef, sausages, and burgers, that replicate the taste, texture, and feel of genuine meat using only plant-based ingredients.

One of Beyond Meat's key advancements in product development is the use of fermentation technology to enhance the nutritional profile and flavor of its products. Through fermentation, Beyond Meat may create complex aromas and textures that are almost identical to those of animal-based foods. This innovative approach not only improves the products' sensory appeal but also meets customer demand for sustainable, healthful food options.

Explanation of the Fermentation Technology Used in Product Development

Beyond Meat's fermentation process uses microorganisms like bacteria and yeast to produce specific proteins and flavors that improve the overall quality of its meat alternatives. One notable example of this is the fermentation method used to produce **pea protein**, a crucial ingredient in many Beyond Meat products. The fermentation process improves the digestibility and flavor profile of the pea protein, making it more appealing to consumers.

Additionally, Beyond Meat mimics the umami taste commonly found in meat by using fermentation to create unique flavor molecules. By carefully selecting specific strains of microorganisms, the company can produce flavor profiles that appeal to meat customers, increasing the enjoyment of plant-based replacements. This biotechnology approach not only improves the taste and texture of Beyond Meat's products but also strengthens the company's commitment to sustainability by reducing the environmental impact of conventional meat production (Fig. 8.3).

Discussion of Consumer Acceptance, Market Trends, and Sustainability Implications

The innovative use of fermentation technology by Beyond Meat has been a key contributor to the economic success of their goods. By faithfully mimicking the flavor and texture of animal meat, Beyond Meat has proven successful in appealing to a broad spectrum of consumers, including flexitarians, vegetarians, and die-hard meat enthusiasts. The company's partnerships with major food chains and retailers have significantly expanded the availability and awareness of its products, contributing to the rapid growth of the plant-based meat sector.

The Beyond Meat fermentation process has important sustainability concerns. By replacing animal-based meat with plant-based meat, Beyond Meat's products can significantly reduce greenhouse gas emissions, land use, and water

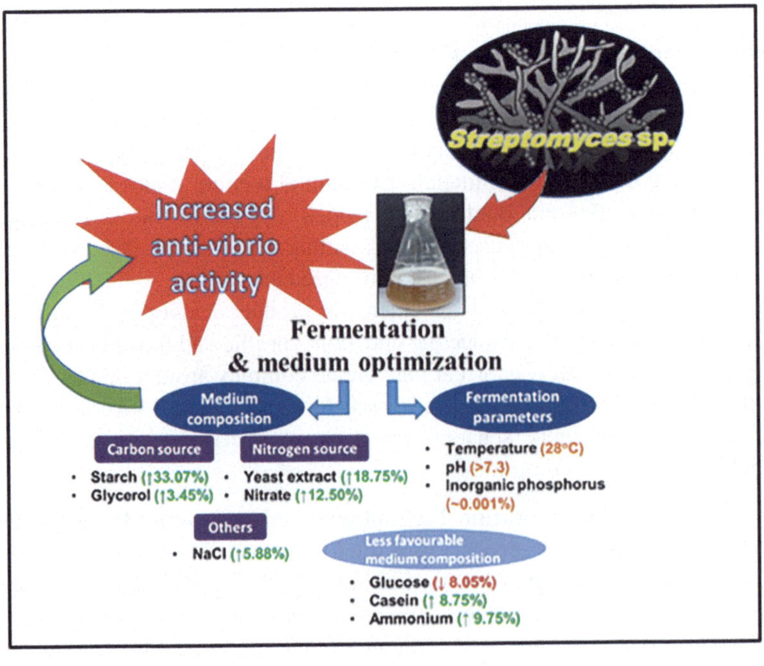

Fig. 8.3 Fermentation process development. (Schäpper et al., 2009)

consumption associated with conventional meat production. Research has shown that, in comparison to the production of cattle, the development of plant-based meats, like Beyond Meat's, can cut greenhouse gas emissions by as much as 90%. Furthermore, fermentation technology minimizes food waste by efficiently converting plant-based ingredients into high-quality protein products.

In conclusion, Beyond Meat's application of fermentation technology is a shining example of the innovative techniques transforming the food industry. By using biotechnology to enhance flavor, texture, and nutritional value, Beyond Meat has successfully created plant-based products that appeal to consumers and promote sustainable eating habits. Thanks to its creative use of fermentation technology and the rising demand from consumers for plant-based alternatives, Beyond Meat is a pioneer in the hunt for delicious and sustainable food solutions.In conclusion, Beyond Meat's application of fermentation technology is a shining example of the innovative techniques transforming the food industry. By using biotechnology to enhance flavor, texture, and nutritional value, Beyond Meat has successfully created plant-based products that appeal to consumers and promote sustainable eating habits. Thanks to its creative use of fermentation technology and the rising demand from consumers for plant-based alternatives, Beyond Meat is a pioneer in the hunt for delicious and sustainable food solutions.

8 Case Study 6: The Circular Economy in Food Processing

Introduction to the Circular Economy Concept

One innovative economic theory that minimizes waste and makes the most of the resources at hand is the circular economy. Unlike the traditional linear economy, which functions on a "take-make-dispose" basis, the circular economy promotes the continuous use of resources by recycling, reusing, and repurposing items. In the food processing sector, this approach seeks to create closed-loop systems that transform food production waste into valuable inputs for subsequent processes. The circular economy reduces the adverse environmental impacts of the food business while simultaneously enhancing its sustainability and resilience.

Application of Circular Economy Principles in Food Processing

The principles of the circular economy can be used in the food processing sector using a number of strategies that prioritize resource efficiency and waste reduction. One of the most well-known applications is **upcycling food waste**, which is converting leftovers of food production into useful items. For example, waste grains from brewing can be recycled into flour or animal feed, and vegetable peels and stems can be utilized to produce snacks or beverages. Companies like Regrained and Winnow Solutions are at the forefront of developing innovative processes that transform food waste into a resource rather than an issue.

Another crucial element of the circular economy in food processing is the use of sustainable packaging. Using biodegradable or compostable packaging materials encourages ethical consumption and lessens plastic waste. Businesses are also investigating closed-loop supply chains, which enable packaging to be recycled and reused, in an attempt to reduce their environmental impact. Initiatives like **looped packaging systems**, which enable customers to return containers for cleaning and reuse, are extending the lifecycle of materials used in food processing.

Economic and Environmental Benefits of Circular Economy Practices
The transition of the food processing industry to a circular economy has significant financial and environmental benefits. By cutting waste and improving resource efficiency, businesses can increase profitability and decrease production costs. For instance, businesses can use food processing by-products to create new revenue streams and boost their overall competitiveness. Studies show that by optimizing resource use and reducing waste disposal expenses, companies that adopt circular economy methods can save up to 30% on expenses.

The circular economy contributes to environmental conservation, pollution reduction, and greenhouse gas emissions. By reusing and recycling items, businesses can lessen their reliance on virgin materials, which often require significant energy and water inputs for extraction and processing. Reducing food waste also decreases the environmental impact of landfilling since food waste breaks down and releases methane, a potent greenhouse gas.

In conclusion, the circular economy is a cutting-edge approach to food production that blends economic feasibility with environmental sustainability. By applying the ideas of the circular economy, the food industry may decrease waste, increase resource efficiency, and create new goods from by-products. Given the growing need for sustainable practices, the circular economy offers food processors a practical way to promote both economic growth and environmental sustainability. Through collaboration and creativity, food supply chain stakeholders can build a more robust and sustainable food system.

9 Conclusion

The implementation of circular economy principles has resulted in a significant shift in food processing toward more sustainable and resource-efficient practices. By rethinking traditional linear models of production and consumption, food processors may increase resource use and minimize waste, which will benefit the environment and the economy. Innovative ways to turn food waste into usable products and the use of eco-friendly packaging show that closed-loop solutions that increase the resilience of the food supply chain are possible.

In order to create an ecosystem that prioritizes sustainability, the circular economy also encourages collaboration amongst various stakeholders, including producers, consumers, and lawmakers. In addition to increasing revenues, companies who recognize the advantages of repurposing leftovers and reducing waste also contribute to resolving the pressing environmental problems that the food industry faces. The transition to a circular economy is not an option; it is a must for food production to remain sustainable in the face of resource shortages and climate change.

In conclusion, the circular economy in food processing promotes environmental stewardship while providing an opportunity for growth and innovation. By embracing these concepts, the food industry can set the standard for a more sustainable future that will benefit not only businesses and consumers but also the environment as a whole. Our continued commitment to circular practices will be necessary to create a resilient food system that can meet the needs of present and future generations.

References

Elhalis, H., See, X. Y., Osen, R., Chin, X. H., & Chow, Y. (2023). Significance of fermentation in plant-based meat analogs: A critical review of nutrition, and safety-related aspects. *Food, 12*(17), 3222.

Flores, A., Wang, X., & Nielsen, D. R. (2019). Recent trends in integrated bioprocesses: Aiding and expanding microbial biofuel/biochemical production. *Current Opinion in Biotechnology, 57*, 82–87.

Schäpper, D., Alam, M. N. H. Z., Szita, N., Eliasson Lantz, A., & Gernaey, K. V. (2009). Application of microbioreactors in fermentation process development: A review. *Analytical and Bioanalytical Chemistry, 395*, 679–695.

Part IV
Enhancing Food Products

Chapter 9
Food Product Development

Zahida Qadeer, Saira Tanweer, Baila Ahmad, Muhammad Ammar Khan, Saadia Zainab, and Aamir Shehzad

1 Food Product Development Process

With an increasing awareness regarding sustainable development goals of the United Nations, more focus has been on the reduction of industrial wastes and production of bioproducts. An industry in which the stages of food processing, distribution, and marketing are interconnected to optimize efficiency, quality, and sustainability may be termed as an integrated food and bioprocessing industry (IFBI). One of the strategies used by such industries is to convert biomass into valuable food products, hereby termed as food bioproducts. Food product development is the process of bringing a new food idea from concept to market. This involves planning, designing, development, scaling-up, commercialization strategies, and product launch. These are explained below.

Z. Qadeer · S. Tanweer (✉) · B. Ahmad · M. A. Khan
Department of Food Science & Technology, Faculty of Agriculture & Environment, The Islamia University of Bahawalpur, Bahawalpur, Punjab, Pakistan
e-mail: saira.tanweer@iub.edu.pk

S. Zainab
College of Food Science and Engineering, Henan University of Technology, Zhengzhou, China

A. Shehzad
UniLaSalle, Université d'Artois,ULR7519-Transformations & Agro-Ressources, Normandie Université, Mont-Saint-Aignan, France

A. Sattar Jatoi (ed.), *Integrated Food and Bioprocessing*,
https://doi.org/10.1007/978-3-031-87209-9_9

1.1 Planning

Initially, the food product development strategic planning starts with product goal setting, conducting market research, and allocating resources for product development (Gumel, 2019). The business objectives of an IFBI may include bioproducts development with reduced effect on environment through enhanced resource efficiency, mainly through low waste generation, low energy utilization, higher quality, and higher yield. In order to align the product goals with the objectives of an IFBI, market research as well as scientific literature assessments are performed to look for insights into consumer perceptions and market trends. For conducting consumer studies, data may be taken regarding demographic factors, interests, behaviors, diseases, etc., and interactions may be derived regarding their behavior towards acceptance of innovative bioproducts. Subsequently, strategic planning is done for the product development according to the market analysis (Chakravarthy, 2019).

1.2 Designing

In the context of IFBI, innovative product design needs to be developed. Designing process involves thinking ideas, developing prototypes, testing quality traits, and designing appealing packaging, while ensuring high efficiency and sustainability. In this process, bioproduct development team may develop and use novel ingredients and processing methods to ensure high resource-efficiency, by repurposing or recycling the food industry wastes. Moreover, compliance to relevant regulations and standards is important. Sensory traits including appearance, taste and texture of new food products are central to product success. In addition, these days health-conscious consumers have great focus on nutrition of food products. Moreover, food industry has prime focus on food quality, safety, and shelf life. Development process of innovative food products may take substantial time. Knowledge of physicochemical properties of various ingredients, bio-additives to optimize the processing condition to achieve develop the prototype of desired product sensory attributes, quality and safety. Furthermore R&D teams must ensure that new products match the compatible costs in the market. In addition, creativity using novel ingredients and processing methods is central to resource efficiency and reduction of bioprocessing wastes. Furthermore, the ingredient sourcing can play central role in cost-effectiveness of the novel products. Finally, sourcing high-quality ingredients is important for maintaining consistency of product attributes (Fig. 9.1).

Packaging is the first thing, which a customer sees, so its design for novel food products requires careful consideration of food product type, target market, cost, and environmental impact. Packaging must be eye-catching for customer, and prevent the product from microbial, enzymatic, mechanical and pest damage. Moreover, it must be convenient for storage, transportation and retail. The packaging of a novel product must also be unique and innovative. Moreover, it must fulfill all the legal

Fig. 9.1 Steps of food product development

labelling requirements and regulatory considerations. However, the R&D teams must be able to create eye-catching designs for the new products, while choosing sustainable materials and balancing customer appeal with cost.

1.3 Development, Scaling-Up and Quality

In IFBI, the development and scale-up process requires lab-scale experimentation and optimization, pilot plant trials, careful process scale-up considerations, and stringent quality control and assurance measures to ensure the production of high-quality, sustainable food products. Initially, small-scale prototypes of the novel bio-product are developed. Subsequently, sensory evaluation, and initial feedback from potential consumers is taken to test the initial feasibility. This follows product and process optimization to match the product traits with consumer expectation, as well as for improvement of process efficiency and sustainably. Once the prototype is final, pilot-plant trials are conducted, which is also known as scaling up process. This involves testing the production process using industrial facilities. Similar steps are repeated regarding consumer feedback and optimization. Throughout this process, special consideration is given to optimizations of process, ingredients, and formulation.

Quality assurance (QA) and quality control (QC) are essential to ensure food safety and quality management in an IFBI. QA is a proactive approach that focuses on preventing defects and ensuring that food products meet established quality standards throughout the production process, establishing training teams, auditing, and risk minimization though preventive measures. QC is a reactive approach that focuses on detecting and correcting defects in finished food products through testing and inspecting products against the defined quality standards and specifications. Since consistent achievement of product quality, safety, shelf life in line to the rules set by regulators is of prime importance, QA and QC work together in the industry to achieve this milestone during the novel food product development, as well as during scaling up.

1.4 Commercialization Strategies and Market Entry

Once the product development, process optimization and market testing are complete, it is time to do proper marketing and regulatory homework prior to product launch. For this purpose, clear plan needs to be developed for when the product will be brought to the market and how it will be introduced. Mostly, companies develop online marketing advertisements, as well as physical product testing campaigns. This not only helps introduction of product to the end user, but also helps collect end user feedback to further improve the product. So, at this stage, it is necessary to develop complete schedule of all these activities. In addition, regulatory compliance is necessary prior to launch of the product according to the local food safety laws, putting the right labels on, and getting any special certifications for the novel product. Compliance to these is the job of QA departments in an IFBI.

1.5 Product Launch and Post-launch Evaluation

Finally, the marketing campaigns are started according to the schedule using planned strategies. Marketing campaigns usually involve free product testing during conferences, at roads, in the marts, etc. This not only gives people the initial taste of the product. If at this stage, product fails to excite the customer, the product fails. Hence, careful monitoring of product reception in the market is important. Companies collect consumer feedback to improve the recipe and try to stay in touch with the loyal consumer using loyalty cards and prizes. The improvement process never ends an IFBI, as it keeps its consumer at the center of its focus to stay competitive in the market.

2 Success and Failure

Novel food products are necessary for the survival of an IFBI. Rutkowski (2022) claimed that about 30–40% of new products fail despite using best strategies of product development and management. The success and failure of non-conventional food products mainly depend on marketing, revenue, and consumer satisfaction. An

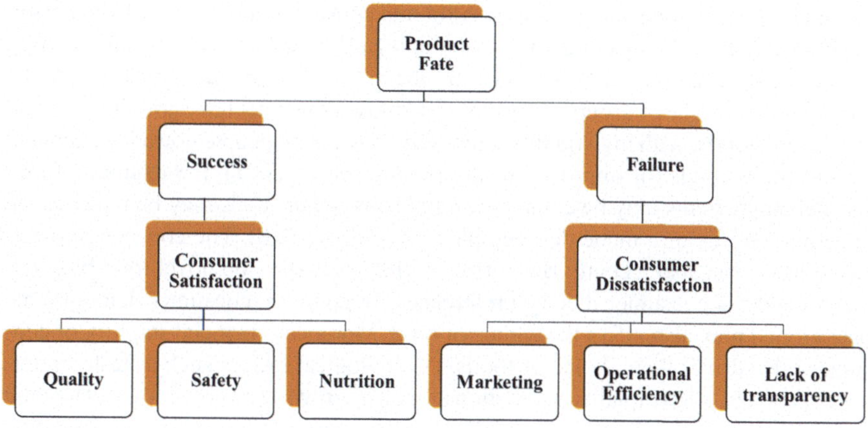

Fig. 9.2 Causes of failure and success of food products

inappropriate marketing campaigns fail to develop a consumer circle and loyalty. Without operational efficiency, a sustainable financial model cannot be developed, which is central to industrial existence. Finally, consumer says is the heart of every innovation, which talks in terms of product quality, safety, nutrition, and price. Quality is the most important criteria, as consumers expect the novel food bioproducts to meet certain standards of taste, texture, and nutritional value. Moreover, consumer health is of prime importance for governments, hence food authorities worldwide force implementation of food laws, failing which a product may be recalled. Recently, nutrition enhancement using bioactive compounds from fruit and vegetable industry remains has gained tremendous interest of researchers, which is adding to the innovation and consumer health arena. Food consumer is now increasingly demanding for organic foods, higher traceability information and reduced animal component incorporation in the products (Azanedo et al., 2020). Finally, environmental impact has become the centerstage of each industry these days, which is not only necessary in the production process but also in the supply chain. The risk sources in supply chains could be environmental, network-related, and organizational, which can be mitigated by avoidance, control, cooperation, and flexibility (Emrouznejad et al., 2023). Machine learning is a novel trend to manage supply chain problems (Yang et al., 2023) (Fig. 9.2).

3 Role of Consumer

Consumer choice and satisfaction play central role in food product innovation in an IFBI. Developing a consumer-based food product requires a comprehensive knowledge of a broad spectrum of requirements (Ratasuk & Charoensukmongkol, 2020). Hence, the innovator needs to know current food preferences, current trends, and factors influencing consumer choices. Although consumer choices vary from person to person, and IFBI may collect information regarding individual consumer dietary requirements (Schmitt, 2012), however, the innovated product must fulfill the dietary needs of a large group of people. These days consumer have growing demand for organic foods, with high quality and safety. Moreover, people nowadays demand for nutritious foods for improved health (Aldaya et al., 2021). For example, foods for diabetic patients may be calorie-free and some other foods may be tailed to the nutritional needs of a niche demographic (Solomon, 2018). The concept of plant-based foods has also recently risen. Hence, IFBI may develop novel food bioproducts to address consumer dietary preferences. In addition, consumer demographic background may also affect the choice of food. Moreover, financial stability of consumers also drives their choice of foods. Psychological factors such as taste preference (Prescott, 2020), quality perception as well as growing awareness about healthy foods (Gehlert & Ward, 2019) may influence the consumer food choices. Even girls in some families may like spicy foods, while boys in the same families may like non-spicy foods. Furthermore, social factors such as cultural background (Schifferstein et al. 2022), social norms (Rozin & Rozin, 2017), as well as family members and friends (Bearden & Etzel, 1982) may also affect the food choices. In addition, innovators must take care of the local cooking techniques, ingredient choices, and preferred flavors to develop a product desirable for consumers. Some Asian countries like umami flavor. The decision-making process can be divided into following phases. Firstly, hunger, cravings, or social incidences create the need for food. At this stage, consumer searches for food, and enquires relevant information such as taste, nutrition, etc. Subsequently, consumer takes reviews from friends and family, and looks for comparative price and brand information. He then enquires for product availability and promotions, then buys the food. After eating it, the consumer rates the quality of food in term of satisfaction, and develops loyalty towards the brand. This leads to repeated purchase of the same food in future (Van Der Merwe et al., 2010). Hence, knowing the decision cycle may help an IFBI to develop food products of consumer choice and need. Involving the consumer in product testing and collecting feedback develops strong relation and loyalty, hence feedback of loyal consumers may help product quality traits improvement. Many companies utilize the approach of co-creation of products, in collaboration with active consumers. This further fosters a sense of ownership among the consumers (Payne et al., 2008).

4 Preference Mapping

Preference mapping in food product development involves several steps including designing studies, conducting trials, establishing evaluation methods, analyzing data, incorporating advanced techniques, utilizing results in product development, and iterative optimization to ensure the product meets consumer expectations. This process uses statistical tools for an accurate mapping of consumer preferences. In an earlier study, Ramón-Canul et al. (2021) used external preference mapping and principal component analysis for selecting a product composition of commercial chocolate flavored milk samples for sugar reduction. In contrast, Perrot et al. (2018) suggested that a multi-market preference mapping could be more useful for coffee mixes, using statistical methods like mixed models, least significant differences, K-Means, principal component analysis, and principal component regression. On the other hand, Varela et al. (2017) explored projective mapping for whole bread and revealed that preference was more weighed than sensory attributes in the process of selection of product. Similarly, Oltman et al. (2016) used external preference mapping using cluster analysis for fresh tomatoes across three stages: appearance, slicing, and consumption and found appearance stage to be associated with highest liking. In addition, Gere et al. (2014) compared various preference mapping methods applied in sensory analysis of flavored kefir products, including categorical principal component analysis, correspondence analysis, and penalty analysis, and reported that the latter provided the most sensitive discrimination among all these methods. However, Ares et al.(2011) used three consumer profiling techniques, including projective mapping, to evaluate the sensory characteristics of orange-flavored powdered drinks, and reported projective mapping providing similar information to external preference mapping. Furthermore, Johansen et al. (2010) used fuzzy clustering in preference mapping and reported that one segment had a linear preference pattern, while the other two had non-linear patterns.

5 Sensory Testing

5.1 Selection of Methods

Sensory evaluation plays an important role in testing the product attributes during the development process. Their scores are considered indicators of consumer preferences. The primary sensory evaluation methods fall into three main categories, viz.; discrimination tests, descriptive analysis, and consumer testing. Discrimination tests aim to detect distinctions or similarities among products or samples. Typical discrimination tests encompass the triangle test, duo-trio test, and paired comparison test. Control sample means the sample against which the quality attributes of the other samples are being tested (Lawless & Heymann, 2010). Descriptive analysis entails trained panelists who systematically assess and articulate the sensory qualities of a product using a predefined vocabulary. Common descriptive analysis methods comprise quantitative descriptive analysis and free-choice profiling (Stone et al., 2008). Consumer testing entails gathering feedback directly from the target demographic, either at home, or at a public place, such as parks in order to assess the real-world acceptance of a trial product (Meiselman, 2013) (Table 9.1).

Table 9.1 Optimization efforts using sensory evaluation data

Stage	Methods	Description
Selection of methods	Discrimination tests	Identify differences between samples
	Descriptive analysis	Trained panelists describe product qualities.
	Consumer testing	Gather feedback from target audience
Testing protocols	Standardization	Consistent testing environment
	Panelist training	Refine sensory perception and understand relevant product attributes.
	Calibration	Maintain panelist consistency
Data analysis	Descriptive statistics	Summarize data
	ANOVA	Identify differences between samples.
	Principal component analysis	Group similar samples based on various factors.
Product optimization	Feedback-driven improvement	Adjust ingredients, additives, and processes based on consumer feedback.
	Balancing qualities	Achieve a product generally accepted by most consumers.
	Targeted audience	Focus on sensory aspects preferred by the target market.

5.2 *Testing Protocols*

Employing dependable sensory testing procedures depends on standardizing procedures as well as on training and calibrating panels in order to validate the accuracy of data gathered during product evaluations. Standardization serves to reduce discrepancies in testing environments (Stone et al., 2020). Panelists need thorough training to refine their sensory perception and grasp the particular attributes pertinent to the products being assessed. Regular calibration sessions are necessary to sustain consistency in panelist performance and detect any deviations (Heymann & Lawless, 2013).

5.3 *Data Analysis*

Data analysis is a complex endeavor. It requires the utilization of statistical methods. In order to interpret the scores of sensory evaluations and establish connections between sensory attributes and consumer preferences, statistical tools are used. Overall impact of treatment is determined using descriptive statistics. These include mean, median, mode, and standard deviation, etc. The differences among various treatments are measured using the Analysis of Variance (ANOVA) test. Complex datasets having hundreds of samples can be distinguished into various groups using Principal Component Analysis. This can be performed using various factors compared against each other, while output is obtained in terms of various treatments arranged in different four quadrants. The closeness in the same quadrant shows similarity of samples, and vice versa. All these approaches help develop a robust framework for interpreting sensory data and help decision-making in product development based on consumer output (Meilgaard et al., 2007).

5.4 Product Optimization

Based upon consumer feedback, the product refinement process is done, by changing various ingredients, additives, and processes conditions. This process is repeated until market acceptance and consumer satisfaction. It is impossible to develop an all-purpose product, and incorporate the suggestions of all the consumers to develop a food product (Yang & Lee, 2019). Hence, a balance is sought among various quality attributes and scores to develop a product acceptable of most of the consumers. The statistical analysis of sensory data of novel products can pinpoint significant sensory components that resonate favorably with the intended audience. For example, in the food sector, striking the correct equilibrium between sweetness and saltiness can greatly influence consumer preference (Meilgaard et al., 2007).

6 Senso-Emotional Optimization

6.1 Emotional Drivers

Besides how food tastes, smells, or looks, the feelings food brings up can greatly affect what a person likes to eat. This can also inspire the making of new food products. Emotions impact behaviors associated with food, such as decision-making regarding food selection. King and Meiselman (2010) characterized emotions as transient, vigorous physiological and mental responses, representing intense mental experiences with heightened hedonic value, encompassing pleasure or displeasure. Meanwhile, the sensory characteristics of food, including visual appeal, aroma, taste, and texture, stand as significant catalysts influencing emotional encounters (Kaneko et al., 2018). Throughout the marketplace, companies widely employ awe-inducing tactics to captivate consumers and advertise their products (Cao et al., 2020). It becomes intriguing to explore the impact of genuine positive and negative emotions in everyday life on how we perceive taste and, consequently, how this might influence our decisions regarding food (Noel & Dando, 2015). Emotional responses are recognized as a distinguishing factor in consumer reactions toward food products. Incorporating the feelings that food brings up in sensory studies can give the food industry a better understanding of what consumers want. Specific sensory qualities are known to correspond to distinct emotions, suggesting a strong link between emotional aspects and sensory experiences. Building upon this connection, this approach suggests that emotions strongly tied to sensory encounters might influence product preference (Mora et al., 2018).

6.2 Emotional Triggers

Emotional eating denotes the inclination to indulge in excessive eating as a means to regulate and alleviate adverse emotions like depression, anxiety, and stress. It signifies a challenge in distinguishing between physiological hunger cues and the use of eating as a strategy to address the negative emotions. Consequently, eating behaviour serves as a method to divert attention or seek relief from unpleasant emotional states (Al-Musharaf, 2020). Individuals experiencing negative moods tend to overeat, favouring energy-dense, appealing foods with mood-enhancing qualities, often associated with their high sugar content (Devonport et al., 2019).

6.3 Psychological Influences

Critical factors influencing general food selection are identified and grouped into distinct categories: internal aspects of food (such as sensory and perceptual attributes), external elements related to food (like available information, social surroundings, and physical environment), individual states (comprising biological characteristics, physiological requirements, psychological aspects, habits, and past experiences), cognitive considerations (involving knowledge, skills, attitudes, preferences, expected outcomes, and personal identity), and socio-cultural aspects (including cultural norms, economic factors, and political influences) (Chen & Antonelli, 2020). Rozin (2006) outlined the impact of various factors—biological (including physiological and evolutionary aspects), psychological (such as preferences and contextual factors), social (pertaining to sociology), and cultural (involving anthropology)—on food selection. The interplay between social and environmental elements, like modelling effects, family environment regarding eating competence, food labelling, taste, appearance, personal food background, habits, and familiarity, alongside psychological determinants such as perceived behavioural control and motivation, has been established to significantly influence eating behaviour (Hardcastle et al., 2015).

6.4 Brand Personality and Consumer Connection

Alignment of brand personality with consumer connection significantly influences brand love and brand loyalty, particularly within the luxury branding domain. This connection underscores the importance of congruence between a consumer's personality and the brand itself. Additionally, the research establishes a link between brand attachment and brand loyalty. Historically, brands sought to secure customer loyalty by emphasizing exceptional quality, and exclusivity. Previous studies have indicated that consumer choices were primarily motivated by the aspiration to boost self-esteem and exhibit social status and prestige (Hamadneh et al., 2021).

6.5 Emotional Loyalty

Customer loyalty represents a committed relationship between customers and a brand or supplier. Customer satisfaction stands out as a key driver in cultivating this loyalty. Amid competitive landscapes, maintaining market share necessitates the creation of a robust brand image. An IFBI can make products more appealing to customers by increasing its advantages, reducing its prices, or by the balance of these two (Farizan et al., 2019).

6.6 Packaging Design and Visual Communication

Packaging serves fundamental roles of safeguarding, containing, providing convenience, and communicating product information (Young et al., 2020). Its design significantly affects consumer perception about the products (Söderlund et al., 2017). While emotional value is not considered a primary function of packaging, it benefits consumers by instilling a sense of making a wise choice and evoking excitement (Peng et al., 2023). For instance, milk in cartons or glass bottles is seen as healthier than that in plastic bottles (Clark et al., 2021). Moreover, visual cues such as color, size, and other physical characteristics increase brand visibility (Yu et al., 2021). Similarly, green or natural-coloured food packaging enhances perceptions of food safety (Marozzo et al., 2020). Moreover, red packaging is linked to tastiness (Luo et al., 2019).

6.7 Sensory Experience and Emotional Resonance

Food companies evoke feelings of awe for positive consumer responses. Aroma, taste, sensation of fat in the mouth, significantly influences the preference and acceptance of foods. Moreover, the enjoyment associated with eating, referred to as hedonics, as well as the emotions associated with food, motivate consumers to either increase or decrease the consumption of a particular food (Noel & Dando, 2015). Moreover, emotions are linked to food experiences. Various senses collaborate to form a cohesive emotional perception of food. There exists a strong connection between taste and smell, as we experience both at same time (Dantec et al., 2021).

6.8 Consumer Engagement and Co-creation

Collaborative co-creation plays a pivotal role in culinary leisure experiences. Engaging customers in co-designing experiential services not only enhances the creation of value but also fosters the development of unforgettable experiences (Batat et al., 2019). This collaborative process has the potential to trigger emotional engagement that contributes to enhancing consumer well-being through co-creation (McColl-Kennedy et al., 2017). The co-creation of food products may be linked to consumer well-being, mainly due to collaborative efforts of consumers and IFBI in innovating processing (Taheri et al., 2021). Hence, consumers may play an important role in transitioning a traditional food system to a sustainable food system for making a successful IFBI (Gomes et al., 2022).

7 Chemical and Instrumental Analysis

Testing food quality is imperative to verification of food properties listed on its label. Proximate analysis is the first step adapted by most of the industries to perform food quality testing. In addition, molecular analysis such as genomic and proteomic techniques can be used to identify the origin of food (Ali et al., 2014). Ultraviolet and visible absorption (UV-Vis) spectrophotometry, nuclear magnetic resonance (LF-NMR), infrared and near-infrared (IR and near-IR) spectroscopies offer quick determination of food components. The volatile aroma compounds are generally determined using gas chromatography–mass spectrometry (GCMS) and electronic nose (Song & Liu, 2018). Atomic Absorption Spectrometry is used for measuring the concentration of various elements in food samples (Md Noh et al., 2020).

Besides, food safety can be measured in terms of food toxicology and microbiology. Heavy metals can originate from feed, drinking water, mineral supplements, and packaging materials (Thakali & MacRae, 2021). Microbial contamination can happen from anywhere (Mead et al., 1999). About 3 million child deaths per year in underdeveloped countries have been associated with microbial sources (Fung et al., 2018).

Rheology represents the fluidic flow of food materials and usually measured with the help of rheometers. The most common attributes which are measured using the rheometer include G, G prime and young's modulus (Zheng, 2019). Texture of food samples is measured in terms of hardness, cohesiveness, gumminess, chewiness, and resilience using a texture profile analyzer (Peleg, 2019). Moreover, Warner-Bratzler shear force test is employed to evaluate the strength with which blade of the apparatus cuts the food sample, and is considered an important indicators of meat tenderness (Kohyama, 2020).

8 Ethics of Food Production and Consumption

With increasing awareness of consumers towards ethics in food production, they now demand the industry to produce food with utmost level of ethics. This also aligns with sustainable development goals (SDGs) of the United Nations (UN). Sustainable sourcing and supply chains, and fair labor practices align with SDG 2 (zero hunger) and SDG 8 (decent work and economic growth). Moreover, food safety and consumer well-being relate with SDG 3 (good health and well-being). Similarly, this SDG also relates with transparency in food industry. Labelling of food items is a legal requirement that helps transparency; and allows consumers informed buying decision. Moreover, reducing the carbon footprint of food production allows meeting the requirements of SDG 12 (responsible consumption and production) and SDG 13 (climate action). Finally, ethical marketing and consumer education are essential for avoiding deceptive practices in marketing a novel food product by an IFBI.

9 Summary

In this chapter, the concept of an integrated food and bioprocessing industry (IFBI) was discussed. The steps of developing novel food bioproducts in an IFBI include planning, designing, development, scaling, up, commercialization and launch. In the success of a novel bioproduct, consumer satisfaction plays the central role, and it is mainly dependent on product quality, safety, and nutrition. Moreover, marketing and operational efficiency also add to the long-term success of a novel food bioproduct in an IFBI. Poor marketing, financial issues, consumer dissatisfaction, and regulations issues could be major hindrances, and can lead to the failure of a novel food bioproduct. Sensory testing is important to measure consumer satisfaction in terms of hedonics. The sensory scores are subjected to statistical evaluation and important information are drawn for improvement of product and process for greater profitability and sustainability. Chemical and instrumental analysis help quality and safety determination of novel food bioproducts in an IFBI. Finally, ethical considerations, including sustainable sourcing, fair labor practices, and transparency, are important for meeting the requirements of SDGs.

References

Aldaya, M. M., Ibañez, F. C., Domínguez-Lacueva, P., Murillo-Arbizu, M. T., Rubio-Varas, M., Soret, B., & Beriain, M. J. (2021). Indicators and recommendations for assessing sustainable healthy diets. *Food, 10*(5), 999.

Ali, M. E., Razzak, M. A., & Hamid, S. B. A. (2014). Multiplex PCR in species authentication: Probability and prospects – A review. *Food Analytical Methods, 7*, 1933–1949.

Al-Musharaf, S. (2020). Prevalence and predictors of emotional eating among healthy young Saudi women during the COVID-19 pandemic. *Nutrients, 12*(10), 2923.
Ares, G., Varela, P., Rado, G., & Giménez, A. (2011). Identifying ideal products using three different consumer profiling methodologies. Comparison with external preference mapping. *Food Quality and Preference, 22*(6), 581–591.
Azanedo, L., Garcia-Garcia, G., Stone, J., & Rahimifard, S. (2020). An overview of current challenges in new food product development. *Sustainability, 12*(8), 3364.
Batat, W., Peter, P. C., Moscato, E. M., Castro, I. A., Chan, S., Chugani, S., & Muldrow, A. (2019). The experiential pleasure of food: A savoring journey to food well-being. *Journal of Business Research, 100*, 392–399.
Bearden, W. O., & Etzel, M. J. (1982). Reference group influence on product and brand purchase decisions. *Journal of Consumer Research, 9*(2), 183–194.
Cao, F., Wang, X., & Wang, Z. (2020). Effects of awe on consumer preferences for healthy versus unhealthy food products. *Journal of Consumer Behaviour, 19*(3), 264–276.
Chakravarthy, G. (2019). A study on marketing strategies for business enhancement. *Think India Journal, 22*(35), 1542–1547.
Chen, P.-J., & Antonelli, M. (2020). Conceptual models of food choice: Influential factors related to foods, individual differences, and society. *Food, 9*(12), 1898.
Clark, E. A., Duncan, S. E., Hamilton, L. M., Bell, M. A., Lahne, J., Gallagher, D. L., & O'Keefe, S. F. (2021). Characterizing consumer emotional response to milk packaging guides packaging material selection. *Food Quality and Preference, 87*, 103984.
Dantec, M., Mantel, M., Lafraire, J., Rouby, C., & Bensafi, M. (2021). On the contribution of the senses to food emotional experience. *Food Quality and Preference, 92*, 104120.
Devonport, T. J., Nicholls, W., & Fullerton, C. (2019). A systematic review of the association between emotions and eating behaviour in normal and overweight adult populations. *Journal of Health Psychology, 24*(1), 3–24.
Emrouznejad, A., Abbasi, S., & Sıcakyüz, Ç. (2023). Supply chain risk management: A content analysis-based review of existing and emerging topics. *Supply Chain Analytics, 3*, 100031.
Farizan, N., Rohman, F., & Hussein, A. S. (2019). The effect of brand identity, brand image, and perceived value on loyalty with customer satisfaction as mediation variable for costumer fresh juice Bintaro. *Jurnal Aplikasi Manajemen, 17*(1), 127–132.
Fung, F., Wang, H.-S., & Menon, S. (2018). Food safety in the 21st century. *Biomedical Journal, 41*(2), 88–95.
Gehlert, S., & Ward, T. S. (2019). Theories of health behavior. In *Handbook of health social work* (pp. 143–163). Wiley.
Gere, A., Kovács, S., Pásztor-Huszár, K., Kókai, Z., & Sipos, L. (2014). Comparison of preference mapping methods: A case study on flavored kefirs. *Journal of Chemometrics, 28*(4), 293–300.
Gomes, G. M., Moreira, N., & Ometto, A. R. (2022). Role of consumer mindsets, behaviour, and influencing factors in circular consumption systems: A systematic review. *Sustainable Production and Consumption, 32*, 1–14.
Gumel, B. I. (2019). The impact of strategic planning on growth of small businesses in Nigeria. *SEISENSE Journal of Management, 2*(1), 69–84.
Hamadneh, S., Hassan, J., Alshurideh, M., Al Kurdi, B., & Aburayya, A. (2021). The effect of brand personality on consumer self-identity: The moderation effect of cultural orientations among British and Chinese consumers. *Journal of Legal, Ethical and Regulatory Issues, 24*, 1.
Hardcastle, S. J., Thøgersen-Ntoumani, C., & Chatzisarantis, N. L. (2015). Food choice and nutrition: A social psychological perspective. *Nutrients, 7*(10), 8712–8715.
Heymann, H., & Lawless, H. T. (2013). *Sensory evaluation of food: Principles and practices*. Springer Science & Business Media.
Johansen, S. B., Hersleth, M., & Næs, T. (2010). A new approach to product set selection and segmentation in preference mapping. *Food Quality and Preference, 21*(2), 188–196.
Kaneko, D., Toet, A., Brouwer, A.-M., Kallen, V., & Van Erp, J. B. (2018). Methods for evaluating emotions evoked by food experiences: A literature review. *Frontiers in Psychology, 9*, 911.

King, S. C., & Meiselman, H. L. (2010). Development of a method to measure consumer emotions associated with foods. *Food Quality and Preference, 21*(2), 168–177.

Kohyama, K. (2020). Food texture–sensory evaluation and instrumental measurement. In *Textural characteristics of world foods* (pp. 1–13). Wiley.

Lawless, H. T., & Heymann, H. (2010). *Sensory evaluation of food: Principles and practices.* Springer Science & Business Media.

Luo, D., Yu, L., Westland, S., & Mahon, N. (2019). The influence of colour and image on consumer purchase intentions of convenience food. *Journal of the International Colour Association, 24*, 11–23.

Marozzo, V., Raimondo, M. A., Miceli, G. N., & Scopelliti, I. (2020). Effects of au naturel packaging colors on willingness to pay for healthy food. *Psychology & Marketing, 37*(7), 913–927.

McColl-Kennedy, J. R., Hogan, S. J., Witell, L., & Snyder, H. (2017). Cocreative customer practices: Effects of health care customer value cocreation practices on well-being. *Journal of Business Research, 70*, 55–66.

Md Noh, M. F., Gunasegavan, R. D.-N., Mustafa Khalid, N., Balasubramaniam, V., Mustar, S., & Abd Rashed, A. (2020). Recent techniques in nutrient analysis for food composition database. *Molecules, 25*(19), 4567.

Mead, P. S., Slutsker, L., Dietz, V., McCaig, L. F., Bresee, J. S., Shapiro, C., et al. (1999). Food-related illness and death in the United States. *Emerging Infectious Diseases, 5*(5), 607–625.

Meilgaard, M., Civille, G., & Carr, B. (2007). *Sensory evaluation techniques* (4th ed.). CRC Press LLC.

Meiselman, H. L. (2013). The future in sensory/consumer research evolving to a better science. *Food Quality and Preference, 27*(2), 208–214.

Mora, M., Urdaneta, E., & Chaya, C. (2018). Emotional response to wine: Sensory properties, age and gender as drivers of consumers' preferences. *Food Quality and Preference, 66*, 19–28.

Noel, C., & Dando, R. (2015). The effect of emotional state on taste perception. *Appetite, 95*, 89–95.

Oltman, A., Yates, M., & Drake, M. (2016). Preference mapping of fresh tomatoes across 3 stages of consumption. *Journal of Food Science, 81*(6), S1495–S1505.

Payne, A. F., Storbacka, K., & Frow, P. (2008). Managing the co-creation of value. *Journal of the Academy of Marketing Science, 36*, 83–96.

Peleg, M. (2019). The instrumental texture profile analysis revisited. *Journal of Texture Studies, 50*(5), 362–368.

Peng, M., Liang, F., Yu, L., & Huang, H. (2023). Smooth or rough? The impact of food packaging design on product healthiness perception. *Food Quality and Preference, 111*, 104970.

Perrot, M., Pineau, N., Antille, N., Moser, M., Lepage, M., Thaler, T., et al. (2018). Use of multi-market preference mapping to design efficient product portfolio. *Food Quality and Preference, 64*, 238–244.

Prescott, J. (2020). Development of food preferences. In *Handbook of eating and drinking: Interdisciplinary perspectives* (pp. 199–217). Springer International Publishing.

Ramón-Canul, L. G., Margarito-Carrizal, D. L., Limón-Rivera, R., Morales-Carrrera, U. A., Rodríguez-Buenfil, I. M., Ramírez-Sucre, M. O., et al. (2021). Technique for order of preference by similarity to ideal solution (TOPSIS) method for the generation of external preference mapping using rapid sensometric techniques. *Journal of the Science of Food and Agriculture, 101*(8), 3298–3307.

Ratasuk, A., & Charoensukmongkol, P. (2020). Does cultural intelligence promote cross-cultural teams' knowledge sharing and innovation in the restaurant business? *Asia-Pacific Journal of Business Administration, 12*(2), 183–203.

Rozin, P. (2006). The integration of biological, social, cultural and psychological influences on food choice. In *The psychology of food choice* (pp. 19–39). Cabi.

Rozin, E., & Rozin, P. (2017). Culinary themes and variations. In *The taste culture reader: Experiencing food and drink* (p. 37).

Rutkowski, I. P. (2022). Success and failure rates of new food and non-food products introduced on the market. *Journal of Marketing and Consumer Behaviour in Emerging Markets, 14*(1), 52–61.

Schifferstein, H. N., Kudrowitz, B. M., & Breuer, C. (2022). Food perception and aesthetics-linking sensory science to culinary practice. *Journal of Culinary Science & Technology, 20*(4), 293–335.
Schmitt, B. (2012). The consumer psychology of brands. *Journal of Consumer Psychology, 22*(1), 7–17.
Söderlund, M., Colliander, J., Karsberg, J., Liljedal, K. T., Modig, E., Rosengren, S., et al. (2017). The allure of the bottle as a package: An assessment of perceived effort in a packaging context. *Journal of Product & Brand Management, 26*(1), 91–100.
Solomon, M. R. (2018). *Consumer behavior: Buying having and being*. Pearson.
Song, H., & Liu, J. (2018). GC-O-MS technique and its applications in food flavor analysis. *Food Research International, 114*, 187–198.
Stone, H., Sidel, J., Oliver, S., Woolsey, A., & Singleton, R. C. (2008). Sensory evaluation by quantitative descriptive analysis. In *Descriptive sensory analysis in practice* (Vol. 28, pp. 23–34).
Stone, H., Bleibaum, R. N., & Thomas, H. A. (2020). *Sensory evaluation practices*. Academic.
Taheri, B., Pourfakhimi, S., Prayag, G., Gannon, M. J., & Finsterwalder, J. (2021). Towards co-created food well-being: Culinary consumption, braggart word-of-mouth and the role of participative co-design, service provider support and C2C interactions. *European Journal of Marketing, 55*(9), 2464–2490.
Thakali, A., & MacRae, J. D. (2021). A review of chemical and microbial contamination in food: What are the threats to a circular food system? *Environmental Research, 194*, 110635.
Van Der Merwe, D., Kempen, E. L., Breedt, S., & De Beer, H. (2010). Food choice: Student consumers' decision-making process regarding food products with limited label information. *International Journal of Consumer Studies, 34*(1), 11–18.
Varela, P., Berget, I., Hersleth, M., Carlehög, M., Asioli, D., & Næs, T. (2017). Projective mapping based on choice or preference: An affective approach to projective mapping. *Food Research International, 100*, 241–251.
Yang, J., & Lee, J. (2019). Application of sensory descriptive analysis and consumer studies to investigate traditional and authentic foods: A review. *Food, 8*(2), 54.
Yang, M., Lim, M. K., Qu, Y., Ni, D., & Xiao, Z. (2023). Supply chain risk management with machine learning technology: A literature review and future research directions. *Computers & Industrial Engineering, 175*, 108859.
Young, E., Mirosa, M., & Bremer, P. (2020). A systematic review of consumer perceptions of smart packaging technologies for food. *Frontiers in Sustainable Food Systems, 4*, 63.
Yu, L., Westland, S., Chen, Y., & Li, Z. (2021). Colour associations and consumer product-colour purchase decisions. *Color Research & Application, 46*(5), 1119–1127.
Zheng, H. (2019). Introduction: Measuring rheological properties of foods. In *Rheology of semi-solid foods* (pp. 3–30). Springer International Publishing.

Chapter 10
Sustainability and Waste Reduction

Muhammad Ammar Khan, Huijuan Yang, Muhammad Waseem, Tahir Mehmood, Faiz-ul-Hassan Shah, Muhammad Saeed Uddin, and Rana Muhammad Bilal

1 Introduction

Consumer demand for sustainable food products is revolutionizing food processing industry towards searching novel processing methods and adaptation of sustainable practices (Otto et al., 2021). Waste produced by the food industry is the biggest hindrance towards achievement of sustainability status in the food industry. Hence, the concept of integrated food and bioprocessing industry (IFBI) has emerged. An IBFI, therefore, aims at controlling waste reduction, not only by reducing its waste production, but also by reprocessing its waste into useful novel bioproducts. Waste is not only generated during the processing, but also throughout the food supply chain, from farm to the fork. Hence, an IFBI focuses on entire food supply chain to achieve the target of sustainability. Sustainability is considered a triple bottom line balancing concept, which means three pillars of sustainability, namely ecological, social, and economic dimensions (Shou et al., 2019). In food processing, this means can be considered triple P, i.e. people (social), the planet (environmental), and profit

M. A. Khan (✉) · M. Waseem · T. Mehmood · Faiz-ul-Hassan Shah
Department of Food Science & Technology, Faculty of Agriculture & Environment, The Islamia University of Bahawalpur, Bahawalpur, Pakistan
e-mail: ammar@iub.edu.pk

H. Yang
College of Standardization, China Jiliang University, Hangzhou, People's Republic of China

M. Saeed Uddin
Department of Human Nutrition & Dietetics, School of Food and Agricultural Sciences, University of Management and Technology, Lahore, Pakistan

R. M. Bilal
Department of Animal Nutrition, Faculty of Animal Production and Technology, Cholistan University of Veterinary & Animal Sciences, Bahawalpur, Pakistan

A. Sattar Jatoi (ed.), *Integrated Food and Bioprocessing*,
https://doi.org/10.1007/978-3-031-87209-9_10

(economic). It is necessary for an IFBI to implement this PPP strategy throughout the food supply chain of its products (Hoang, 2021).

In the achievement of its goal of sustainability throughout the entire food supply chain, an IFBI may face challenges related to energy efficiency (Menon et al., 2020), lowering greenhouse gas emissions (da Silva et al., 2021), initial investment costs, resistance to change, educating stakeholders, social responsibility, compliance with regulations, logistical issues, and food safety. Overcoming these challenges, however, will be rewarding (Rabadán et al., 2019), as implementation of a sustainable system in the food supply chain will not only enhance the efficiency for all components of an IFBI, but also increase its economic advantages (Arshad et al., 2021). These include reduced operational costs through resource efficiency, improved brand reputation and customer loyalty, access to new markets and discerning consumers seeking sustainable products, market competitiveness, and enhanced resilience to regulatory changes and supply chain disruptions. Additionally, these practices catalyze innovation and product differentiation, thereby driving a competitive advantage and long-term profitability (Sayas-Barberá et al., 2022). Finally, sustainability initiatives can promote employee engagement, and contribute to a more stable and sustainable future for the company and the food industry as a whole. The tools which may help an IFBI to achieve the destination of sustainability include responsible sourcing (Kamgang et al., 2023), waste reduction (Gracia & Gomez, 2020), water conservation (Eslami et al., 2023), ethical labor practices, humane handling of food animals/birds (Sardar et al., 2023), adapting sustainable transportation practices (Jasim et al., 2022), and using sustainable packaging (Tyagi et al., 2021).

Food processing operations can generate a significant amount of waste, including organic by-products. These are often thrown to landfills or sent to incineration. The food industrial waste is a waste of food inputs often food ingredients, for which the food industry has already paid substantial cost (Belhadi et al., 2022). Moreover, companies have to pay additional cost for their disposal. These resources are not only valuable but often limited (Yadav et al., 2020). They may have a significant environmental impact, as well. Circular economy is the concept in which the wasted resources are recycled, reused, and repurposed (Yadav et al., 2020). This concept is known as value-addition in food processing. Government agencies are promulgating regulations and industry standards increasingly to control emissions and increase waste management to prevent environmental pollution. So, waste reduction has become necessary in food processing industry. Efficient waste management strategies play crucial role in the sustainability of food processing. Optimization of the resources in the food industry is the way to go. Minimization can be achieved through efficient process design and utilization of waste-to-value strategies (Feng et al., 2021). Considerations for selecting the most appropriate waste reduction techniques in bioprocessing include the type and volume of waste generated, available technologies, and compatibility with existing processes. Waste reduction improves the overall efficiency and productivity of bioprocessing operations by minimizing resource wastage, reducing process downtime, and lowering costs associated with waste disposal. Bioprocessing companies can measure and track their progress in waste reduction through metrics like waste generation rates, recycling percentages, and cost savings associated with waste management. Moreover, food processing industry can develop innovative ways to repurpose by-products, which give them opportunity to create new revenue streams, as well as a viable tool to reduce operational costs (Feng et al., 2021). By lowering expenses associated with the disposal and treatment, companies can improve their productivity. These can substantially mitigate the environmental impact of bioprocessing waste, particularly contributing to reduced pollution and conservation of natural resources (Sosa-Martínez et al., 2023). Moreover, adopting to waste reduction protocols will ensure compliance with the environment related regulations of governments, which are tightening day-by-day. The food industry which implements waste-reduction strategies also fosters positive relationships with its customers, investors, and the local community, which enhances its brand image and builds trust by showcasing a dedication to environmental stewardship.

The primary objectives of this chapter are, therefore, to elucidate the pivotal role of sustainability in food processing, explore actionable insights, strategies and technologies, as well as spotlight the significance of waste reduction for achieving sustainability in food processing.

2 Sustainable Development Goals

Sustainability and waste reduction in an IFBI correlate with sustainable development goals (SDGs) of the United Nations (UN). Particularly, the objective of SDG 12 is sustainable production and consumption. Reducing food wastage, adopting eco-friendly packaging solutions, and promoting sustainable sourcing and procurement all contribute to meeting SDG 12 in an IFBI (Djekic et al., 2021). An IFBI can also directly or indirectly contribute to the achievement of the objectives of few other SDGs. For example, an IFBI ensures an efficient food manufacturing mainly by low generation of food wastes can indirectly contribute to the achievement of SDG 2, which aims to end hunger, achieve food security, and promote sustainable agriculture (Mensi & Udenigwe, 2021). In addition, an IFBI also ensures implementation of QA and QC systems in its processing facility to ensure maintenance of utmost quality and safety of the food products. In this way, an IFBI meets with the requirements of SDG 3, which targets ensuring healthy lives and promoting well-being for all at all ages (Fernandez, 2020). Higher magnitude of water is wasted in food processing industry during cleaning of the production surfaces and for the removal of food residues to prevent contamination. This water contains a lot of bioactive compounds mainly derived from the food being processed. When such water goes outside the food processing premises, it can cause measurable harmful impacts to environment. An IBFI can control its environmental impact by controlling the amount of its effluents, also by treating the water before releasing it outside into natural water reservoirs. In this way, an IFBI indirectly contributes to the SDG 6, which focuses on ensuring access of clean water to all (Baskaran, 2022). Lastly, SDG 13 emphasizes climate action, a realm where sustainable practices help mitigate greenhouse gas emissions, and contribute to climate resilience (Küfeoğlu, 2022).

3 Sourcing and Procurement

Sourcing refers to supplier selection and contracting, whereas procurement means ordering, receiving, inspection, payment, and record-keeping. Both these activities are important for ensuring the consistent quality and integrity of food products. An IFBI should have a clear sustainability criterion for ingredient sourcing. Ingredient sourcing and procurement are performed based on a number of criteria, such as quality conformance, ability of a supplier to provide consistent supply of the ingredients, price, distance to be traveled, time for consignment to reach the IFBI, alternatives, etc. Usually, two to three suppliers are selected to ensure consistent competition and avoiding any break of supply in case of any miscommunication. However, in an IFBI, these activities may incur huge environmental impact, and if any of these practices is not sustainable, a myriad of challenges can arise. These include environmental degradation through overexploitation of natural resources. Moreover, they may be vulnerable to market fluctuations and regulatory changes. Hence, an IFBI should minimize its environmental impact judicious selection of raw materials and ingredients.

Sourcing and procurement in an IFBI involve a several strategies to achieve the objective of sustainability. One of the most important approach in this regard is performing sourcing and procurement activities from local areas. It promotes strengthening of local communities, as well as reduces carbon emissions associated with transportation (Coley et al., 2009). Moreover, this approach is suitable for promoting fresher and more traceable products. On the other hand, long and complex supply chains results in inefficiencies, increased food miles, and higher disruptions. Moreover, an IFBI prioritizes suppliers that exhibit implementation of high ethical standards. In addition, IFBI may prefer sourcing the ingredients that have sustainable certifications, such as organic, Fair Trade, or Rainforest Alliance, etc. Furthermore, food supply chain may result to food wastage and safety issues, hence optimizing efforts can increase sustainability of an IFBI. Furthermore, using novel technology can help the cause of sustainability in an IFBI. For example, blockchain and IoT sensors can increase transparency and accountability in the supply chain, while data analytics and artificial intelligence tools increase efficiency of decision making in sourcing and procurement (Sunny et al., 2020). Since regular monitoring enhances overall supplier performance, regular supplier audits may be conducted to check the compliance to sustainability standards in an IFBI. Moreover, collaboration efforts with the suppliers enhance traceability and reduce resource wastage. Finally, mock traceability practice in an IFBI may be used to verify the efficiency of the system at a defined frequency.

4 Waste Minimization and Recycling

Food wastes contain considerable quantities of bioactive components, which are harmful for the environment. Most of these are thrown into landfills, and upon oxidation, they produce obnoxious gases and chemicals that pollute the air. Moreover, they contain nutritive components that harbor bacterial growth, hence pose serious health risks and cause substantial damage to environment. Governments and non-governmental agencies have put special emphasis on the need to control food industry wastages. Many challenges can hinder the efforts of the food industry towards waste minimization. Improper waste management may lead to increased disposal costs. Effective waste recycling strategies may reduce carbon footprint of the industry, hence contributes to a reduction in environmental degradation (Bhatia et al., 2023). Since recycling can increase the resources, it bolsters the economic sustainability of the food industry. The efforts to minimize waste start from implementation of waste stream analysis, which helps identify specific areas for improvement. Subsequently, advanced waste reduction technologies may be implemented, and clear performance metrics may be established.

One of the primary areas of improvement is where the waste is produced. If the processing machine is not performing optimally, repair and fixing it for optimal output is the first step. However, if the machine is no longer sustainable, an equipment upgrade is desirable. Sometimes processing steps lead to wastage during food processing. In such cases, process adjustments and technological upgrades are required. Optimization may be achieved by adjusting various combinations of process temperatures, times, pressures, or other processing conditions. Moreover, many novel additives can also be useful to enhance processing output. However, some wastes are inevitable, which require recycling. Organic wastes, for example, can be converted into valuable resources (Suri et al., 2021). Innovative technologies hold significant promise for revolutionizing waste minimization efforts in the food industry. Among the sustainable approaches is the closed-loop model, which emphasizes the reduction, reuse, and recycling of materials, serves as a cornerstone for sustainable waste management in the food industry. Moreover, adopting circular economy principles promotes the repurposing of waste materials, ensuring a more sustainable and resource-efficient operation (Yadav et al., 2020). For instance, the wastewater can be recycled for non-drinking purposes, while wasted packaging material can be converted into packaging material for other uses. Traditionally, single-use packaging has been a serious environmental concern. However, now it is possible to develop primary packaging from biodegradable resources, such as fruits and vegetable pulp. Similar novel technologies can contribute to reduce the load of single-use packaging on the planet. However, recycling strategies may also be performing low, so they may also be needed to be optimized. Particularly, inefficient waste segregation methods can lead to suboptimal recycling performance. The wasted feathers, bones, and inedible portions from chicken processing industry can be cooked to make protein powder to be added as broiler feed supplement. The bioprocessing wastes obtained from the juice extraction can be converted into broiler feed, or packaging material.

Waste minimization not only contributes to increased profitability but also enhances the environmental sustainability for the food industry. Additionally, exploring opportunities for by-product valorization can pave the way to new revenue streams while simultaneously reducing waste volumes. Continued research and development in waste-to-value technologies are pivotal for effective waste management and recycling. Several integrated approaches are being implemented in food production, processing, and packaging industries which have been significantly contributing to sustainability and waste reduction efforts. Total Quality Management (TQM) emphasizes continuous improvement and waste reduction through process optimization, quality control, and employee involvement. Lean manufacturing principles promote efficiency and sustainability by reducing wastes, including excess materials, energy, and time. Just-in-Time (JIT) inventory management minimizes wastes by producing or ordering only what is needed, when it is needed. Integrated Pest Management (IPM) ensures food safety by controlling pests, hence reduces food related wastages. Life Cycle Assessment (LCA) identifies areas for waste reduction by evaluating the entire life cycle of a food product from production to disposal. Circular economy principles promote recycling and repurposing of waste materials (Yadav et al., 2020).

5 Process Optimization for Sustainability

Reliance on energy-intensive processing methods in the food industry as well as the use of non-renewable resources hinders the sustainability in processing operations. Consequently, optimization of existing food processing operations may be approached in a structured manner. In this regard, life cycle assessments (LCA) of existing food processing steps may be conducted to identify environmental hotspots, followed by implementation of lean manufacturing practices (LMP). Moreover, eco-design principles may also be integrated into product and process development. LCA evaluates of the environmental impact of a food product from farm to fork. In the first step of LCA, it is essential to clearly define the goal of the LCA as well as establish its scope. Subsequently, it is important to assess the LCA of each stage of each raw material and food product in the inventory. Thirdly, the impact of each process and product on the greenhouse gas emissions, water use, land use, and other relevant environmental impact indicators will be assessed. In the next step, we need to interpret the results into useful conclusions for identification of opportunities for environmental improvement. In the process of LCA, collection of accurate and reliable data for all stages of a food product's life cycle may be challenging. Due to higher diversity of food products, depending upon a wide range of ingredients and production conditions, conducting LCAs for each product can be resource-intensive. LMP systematically helps in reducing the waste and improving efficiency in food production processes, hence serves as a cornerstone for the sustainable process optimization. The five core principles of LMP include specification of the value, mapping of the value stream, optimization of the flow, pull of the production system, and in last, the perfection (Leite & Vieira, 2015). Firstly, the quality, safety, and organoleptic value of the food product create the value of a product with respect to consumers' point of view. Subsequently, a visual representation from raw material sourcing to the final packaging represents the value stream. Thirdly, the optimization of process flow to ensure highest efficacy, lowest wastage, and lowest delays is important. In the next step, the avoiding excess production beyond customer demands and storage capacity keeping in view product perishability represents the pull in the LMP. Finally, the continuously effort for improvement through regular mock tests, consumer feedbacks, internal and external processes audits, represent the last step of LMP. During the implementation of LMP, the food processor may face challenges regarding the seasonal availability of certain raw materials which may not only affect the production schedules but also cause effect on non-uniform product quality, that warrants the need for flexibility in the strategies. Moreover, optimization of complex processing steps and trial of novel ingredients may require utilization of statistical tools. These days AI-based tools have bombarded every arena of science, and they have high scope of adaptation in the food processing sector as well. Additionally, adopting green processing techniques, such as energy-efficient equipment and environment-friendly production methods, etc., contributes to a more sustainable operation; however, they may also require huge initial capital. Moreover, the existing processes may be redesigned to make them eco-friendly to reduce their environmental impact.

6 Packaging

Inappropriate packaging leads to product spoilage, safety concerns, product integrity, and excessive wastage of food products, which is a serious concern for food industry due to economic losses and environmental sustainability issues. In this regard, adoption of sustainable packaging within the food processing sector can be helpful for minimizing environmental footprint, as well as to meet the demands of eco-conscious consumers for food products. It is prudent to conduct LCA of the packaging material used for food packaging. The most important sustainability issues exist with existing packaging practices within the food industry is related to reliance on single-use plastics and non-recyclable packaging materials (Walker et al., 2021). Moreover, excess packaging waste due to non-compatible design is also an issue. Innovative packaging designs with traditional packaging material may increase material use efficiency and reduce excess packaging waste. Moreover, optimization of packing machines may also reduce the processing wastages related to packaging. A structured approach to packaging for sustainability may include steps like: the assessment of sustainable packaging materials, design optimization for material efficiency, and the implementation of innovative packaging technologies. Recycling of wasted packaging material may contribute to circular economy (Yadav et al., 2020). The novel trend towards sustainable packaging approaches include use of biodegradable plastics, compostable packaging, and recycled materials for development and design of the food packaging material. Edible packaging material is also gaining much popularity at various scientific forums. Moreover, many smart packaging technologies, such as active packaging and intelligent labels have been scientifically trialed to extend shelf lives of food products, hence reduce food wastage (Pandey et al., 2022). Furthermore, the integration of novel technologies in the existing processing steps is a major challenge, which need collaborations of scientific forums, such as universities, research organizations, packaging industry, and food processing industry.

7 Transportation

High carbon emissions may be the direct outcome of inefficient transportation practices in the food industry. Food supply chain is a complex process, which may vary from cold chain, frozen chain, dry chain, live-animal transport, live-bird-transport, fish aquarium transport, etc. Particularly, the management of temperature-sensitive products during transit necessitates innovative solutions for refrigerated transportation, which increases the consumption of resources, hence increase greenhouse gas emissions. This supply is susceptible to multiple vulnerabilities at various stages from the source of procurement, during journey, as well as retail. It is, therefore, essential to carefully assess all the sites of vulnerabilities, where transport-related losses may occur in the foods, particularly related to food, food quality, food safety, packaging, etc. Moreover, the concept of food miles is another environmental effect which is needed to be kept in mind prior to designing efficient food supply chain strategies (Li et al., 2022). Traditional transportation vehicles which mainly rely on fossil fuel have high carbon emissions. Firstly, optimization of food packaging size and package weights may reduce the product load on the transportation vehicles. Moreover, route optimization and load consolidation practices can help to reduce fuel consumption and emissions. Furthermore, audits may be conducted to identify areas for improvement, as well as for establishment of performance metrices. Additionally, drivers may be trained to promote fuel-efficient driving habits. Instead of road transportation, the utilization of rail and water transportation is more eco-friendly, however, these can only be used for longer distances. Moreover, wear-and-tear of old vehicles may further reduce the fuel consumption efficiency of the vehicles. One of the eco-friendly approaches is to use electric and hybrid vehicles for the transportation of food items. In order to reduce food miles, local sourcing and selling may be promoted.

8 Water Usage and Energy Efficiency

Resource management is important for sustainability in an IFBI. In case of inefficient management of resources, process and product costs could be high. Moreover, wastage of ingredients can increase. Water is used in food industry in huge amount. It is required for cleaning of all production surfaces, as well as walls, floors, and safety gear of workers in order to prevent contamination. Huge amount of energy is consumed for acquiring underground water. Moreover, when this water goes out of the food processing facility, it contains the bioactive compounds derived from food items processed using them. The water containing bioactive compounds is hazardous for the environment. Hence, water must be cleaned and recycled prior to throw out of the processing facility. In addition, energy usage in extraction of water also adds to operational costs, which ultimately leads to higher carbon emissions. Hence, optimization of energy resources and water usage are important for the sustainability of an IFBI. Moreover, the wasted ingredients and products in the industry also do not pay back anything in terms of electricity bill or company income. Hence, judicial use of water during processing, recycling and treatment of used water for secondary uses, and the utilization of food industry by-products for secondary purposes can save the important energy sources used for extraction of underground water and purchase of food ingredients.

A lot of wear and tear can happen in industrial implements and tools, repairing which can add to the production cost. Particularly, old product manufacturing equipment, such as in snacks industries may have higher wear and team, which may require heavy repair and maintenance work. Higher wear and tear results in higher energy consumption. Moreover, it can reduce the production efficiency. Hence, regular maintenance can ensure functioning of equipment at peak efficiency. Such costs can be reduced by equipment upgrades or by adoption of advance technologies. However, economic viability may be checked prior to equipment upgradation and integrating renewable energy resources in the existing systems. These may have high initial cost, but may be economical in the long run.

During industrial processing, a lot of heat is generated, which often remains unused. This heat usually increases environmental temperature of controlled environment of food processing premises, which is often undesirable. Implementing heat recovery systems in the processing facilities can help capturing and reusing heat generated during processing.

Integration of renewable energy sources in the existing energy production and utilization in an IFBI may cut energy costs and carbon footprints. Although, these could be expensive, but replacing old non-energy efficient processing equipment with newer green and energy-efficient machinery and appliances can payback soon by lower energy consumption bills. Renewable energy integration stands as a pivotal component of sustainable energy management in the food industry. This may involve the installation of solar panels, wind turbines, or biomass energy systems. By harnessing clean and renewable sources of energy, companies can significantly reduce their environmental impact and dependence on non-renewable resources.

Process optimization is an important tool to reduce wastages and increase process efficiency (Singh et al., 2023). Statistical tools can be used for optimizing the processes in the food industry to minimize energy consumption. In this regard, a structured approach to energy efficiency and resource management may include conducting energy audits to identify areas for improvement, implementing energy-efficient technologies, and establishing clear performance metrics. Systems such as ISO 9001 and ISO 22000 focus on effect quality management and food safety in food production (Agus et al., 2020). They help the management achieve sustainability via reducing wastages by keeping record and taking corrective actions. In light of these systems, companies actively engage employees in energy conservation efforts through training and awareness programs.

9 Regulatory Compliance and Certification

Government agencies and international bodies continuously create regulations and raise awareness regarding sustainability and waste management in the food processing industry. They set standards, and provide incentives and impose penalties for non-compliance to those standard. Consumer, nowadays, has increased tremendous awareness about these regulations, which are particularly related to food quality, safety, and sustainability. Hence, in order to gain greater market access and consumer trust, an IFBI need to comply with local and international regulations. The main areas of concern are product integrity, waste management, emissions control, and ethical sourcing. To address these challenges, an IFBI need to follow a structured approach to regulatory compliance and certifications. This involves several key steps, including conducting thorough assessments of applicable regulations, implementing robust management systems, engaging in continuous monitoring, corrective actions, and continuous improvement.

An IFBI may go for various certifications to achieve sustainability goals. International Organization for Standardization (ISO) has several environmental management systems including ISO 14001, 14,004, 14,006, 14,064 and ISO 5000. ISO 14001 is a systematic approach to protect the environment in an IFBI. Its compliance mainly helps an IFBI to manage water and materials use, waste disposal, control of emissions, and energy conservation. In addition, ISO 14004 provides guidelines for establishing ISO 14001. Moreover, ISO 14006 also uses ISO 14001 provides guidelines, but mainly focuses on design of processes and processing facility. ISO 14064 provides a framework for measuring greenhouse gas emissions. Finally, ISO 5000 targets energy use efficiency through optimization of design, development and processing practices.

Hazard Analysis and Critical Control Points (HACCP) is a systematic approach to food safety. It addresses biological, chemical, and physical hazards in an IFBI. Key principles of HACCP include conducting hazard analysis, determination of critical control points, establishment of critical limits, establishment of monitoring procedures, taking corrective actions, verification and validation, and record

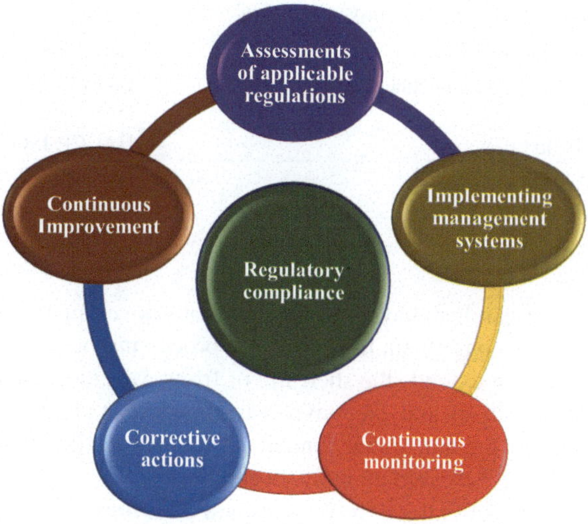

Fig. 10.1 A structured approach to regulatory compliance

keeping. ISO 22000 certification focuses both on food safety by incorporating the requirements of HACCP, as well as quality management system which is ISO 9001 (Agus et al., 2020). Food Safety System Certification (FSSC) 22,000 is an upgradation of ISO 22000, and it holds more international acceptance than ISO 22000. It is recognized by Global Food Safety Initiative (GFSI) (Fig. 10.1 and Table 10.1).

With growing awareness of consumers towards health foods, organic foods certifications have gained great interest. International Federation of Organic Agriculture Movements (IFOAM) is a global organization which establishes standards for qualifying foods as organic (Willer et al., 2024). In addition, various countries may have their own standards and certifications for organic status of foods, such as USDA Organic certification in US and EU Organic in EU countries.

In addition, fair trade certification indicates compliance with environmental, economic, and social standards. It mainly complies with ISO 17065 standard. Its purposes are empowering producer, promoting equitable trade practices, fair wages, and humane working conditions for agricultural laborers. Hence, an IFBI can follow the route of sustainability through achieving all the above certifications.

10 Future Trends and Innovations

Traditional food processing methods often employ excessive use of energy, produces large quantities of processing wastes, require huge water inputs, and are resource inefficient. However, the future of sustainability in the food processing industry is poised for significant advancements through emerging technologies and innovative trends.

Table 10.1 Sustainability and quality certifications for IFBIs

Category	Certifications
Environmental management	ISO 14001, ISO 14004, ISO 14006, ISO 14064, ISO 5000
Food safety	HACCP, ISO 22000, FSSC 22000
Organic	IFOAM, USDA Organic, EU Organic
Fair trade	Fairtrade Certification (complies with ISO 17065)

Researchers have developed various novel innovative food processing technologies. Among them, high-pressure processing has gained the most industrial adoptability. It extends the shelf life of fruits, vegetables and juices without the need for preservatives and excessive heat (Sreedevi et al., 2020). It preserves foods without altering the nutritional contents of foods. Its utilization in meat processing industry has also increased globally. In addition, pulse electric field has also been utilized for increase of extraction of juices and oils from fruits and vegetables (Brito & Silva, 2024). It also minimizes processing wastes. Likewise, membrane filtration systems that have been utilized for separation and purification of components like proteins and enzymes, also reduce water wastage and increase resource use efficiency. Another innovative processing method is closed-loop systems, which is used in the processing of foods. It recirculates and reuses process water, minimizes water consumption and reduces wastewater generation (Rufí-Salís et al., 2020). Moreover, the development of plant-based and alternative proteins aims to reduce dependency on traditional animal agriculture. Circular economy models are being designed to create systems that not only minimize waste but also maximize resource efficiency (Yadav et al., 2020). Additionally, blockchain technology is being harnessed to enhance transparency and accountability within the supply chain (Sunny et al., 2020). Innovations in advanced packaging materials, such as eco-friendly and biodegradable solutions, are set to revolutionize the industry. Waste valorization technologies can transform food waste into valuable products (Roy et al., 2023). Furthermore, the tracking of carbon footprints is helping mitigation of emissions throughout the production process. Finally, artificial intelligence tools are gaining tremendous growth these days, which can be helpful in optimization of food production. In short, the future of IFBIs seems promising and environmentally sustainable (Fig. 10.2).

11 Conclusion

Achievement of sustainability is the basic goal of an integrated food bioprocessing industry (IFBI). Several SDGs of UN are directly or indirectly related to sustainability. SDG 12 addresses the waste reduction, eco-packaging, and sustainable sourcing, while SDGs 2, 3, 6, 9, and 13 targets the hunger, health, water, technology,

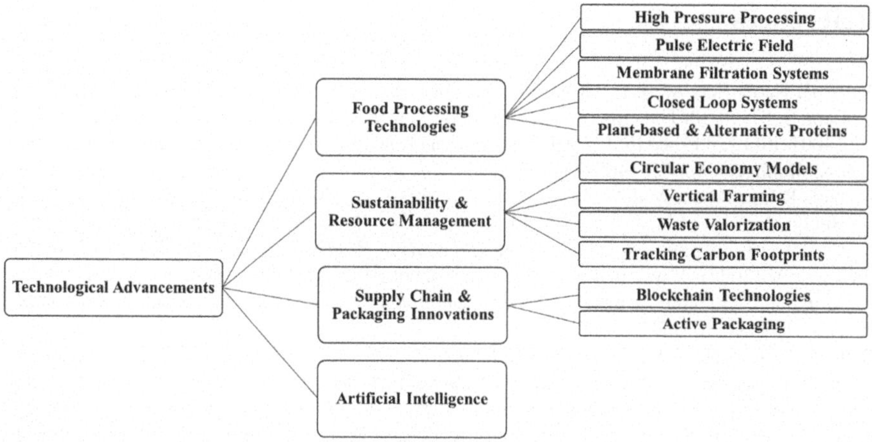

Fig. 10.2 Emerging technologies and innovations for sustainable and efficient food processing industry

and climate action, respectively. An IFBI can achieve its goal of sustainability through making effort towards achieving these SDGs. Optimization of the sourcing and procurement practices are first steps towards the destiny of an IFBI. Next step could be minimization of waste and their recycling. Statistical tools and AI tools may be used for process optimization through collecting process data and readjustment of produce recipes and processing conditions to achieve higher process efficiency. Novel trends in packing, such as active packaging, vacuum packaging, controlled atmospheric packaging, modified atmospheric packaging, etc. can improve product shelf life, hence contribute towards waste reduction. Moreover, use of plant-based recyclable materials for making packing materials for food products also adds to sustainability. Efficient transportation practices in the food industry help to mitigate the challenges associated with carbon emissions throughout the food supply chain. These include refrigerated transportation, optimizing packaging size, route planning, load consolidation, driver training, and transitioning to electric or hybrid vehicles. Moreover, reducing food miles could also be a useful approach to reduce carbon footprints due to transportation. In a food industry, water usage is major energy consumer, hence water use could be limited. Moreover, it is also a source of environmental contamination, hence it must be treated prior to throwing out of the industry. Similarly, processing equipment can wear and tear with age, and consume higher energy. So, regular maintenance and periodic upgrades can be useful in the long term. Compliance to government regulations and international standards such as ISO, HACCP, FSSC 22000, organic, and fair-trade certifications, can be instrumental in achieving sustainability status in an IFBI. Finally, researchers are exploring novel sustainable processing methods, which can be used for efficient resource management, reducing wastage, increasing shelf life, and improving process efficiency in an IFBI.

References

Agus, P., Ratna Setyowati, P., Arman, H. A., Masduki, A., Innocentius, B., Priyono Budi, S., & Otta, B. S. (2020). The effect of implementation integrated management system ISO 9001, ISO 14001, ISO 22000 and ISO 45001 on Indonesian food industries performance. *Test Engineering and Management., 82*, 14054–14069.

Arshad, R. N., Abdul-Malek, Z., Roobab, U., Munir, M. A., Naderipour, A., Qureshi, M. I., Bekhit, A. E.-D., Liu, Z.-W., & Aadil, R. M. (2021). Pulsed electric field: A potential alternative towards a sustainable food processing. *Trends in Food Science & Technology., 111*, 43–54.

Baskaran, V. (2022). A systematic review on the role of geographical information systems in monitoring and achieving sustainable development goal 6: Clean water and sanitation. *Sustainable Development, 30*, 1417–1425.

Belhadi, A., Kamble, S. S., Venkatesh, M., Jabbour, C. J. C., & Benkhati, I. (2022). Building supply chain resilience and efficiency through additive manufacturing: An ambidextrous perspective on the dynamic capability view. *International Journal of Production Economics, 249*, 108516.

Bhatia, L., Jha, H., Sarkar, T., & Sarangi, P. K. (2023). Food waste utilization for reducing carbon footprints towards sustainable and cleaner environment: a review. *International Journal of Environmental Research and Public Health, 20*, 2318.

Brito, I. P. C., & Silva, E. K. (2024). Pulsed electric field technology in vegetable and fruit juice processing: A review. *Food Research International*, 114207.

Coley, D., Howard, M., & Winter, M. (2009). Local food, food miles and carbon emissions: A comparison of farm shop and mass distribution approaches. *Food Policy, 34*, 150–155.

da Silva, J. T., Garzillo, J. M. F., Rauber, F., Kluczkovski, A., Rivera, X. S., da Cruz, G. L., Frankowska, A., Martins, C. A., da Costa Louzada, M. L., & Monteiro, C. A. (2021). Greenhouse gas emissions, water footprint, and ecological footprint of food purchases according to their degree of processing in Brazilian metropolitan areas: A time-series study from 1987 to 2018. *The Lancet Planetary Health, 5*, e775–e785.

Djekic, I., Batlle-Bayer, L., Bala, A., Fullana-i-Palmer, P., & Jambrak, A. R. (2021). Role of the food supply chain stakeholders in achieving UN SDGs. *Sustainability, 13*, 9095.

Eslami, E., Abdurrahman, E., Ferrari, G., & Pataro, G. (2023). Enhancing resource efficiency and sustainability in tomato processing: A comprehensive review. *Journal of Cleaner Production, 425*, 138996.

Feng, J.-Y., Wang, R., Thakur, K., Ni, Z.-J., Zhu, Y.-Y., Hu, F., Zhang, J.-G., & Wei, Z.-J. (2021). Evolution of okara from waste to value added food ingredient: an account of its bio-valorization for improved nutritional and functional effects. *Trends in Food Science & Technology, 116*, 669–680.

Fernandez, R. M. (2020). SDG3 good health and well-being: Integration and connection with other SDGs. In *Good health and well-being* (pp. 629–636). Springer International Publishing.

Gracia, A., & Gomez, M. I. (2020). Food sustainability and waste reduction in Spain: Consumer preferences for local, suboptimal, and/or unwashed fresh food products. *Sustainability, 12*, 4148.

Hoang, V. (2021). Modern short food supply chain, good agricultural practices, and sustainability: a conceptual framework and case study in Vietnam. *Agronomy, 11*, 2408.

Jasim, N. I., Kasim, H., & Mahmoud, M. A. (2022). Towards the development of smart and sustainable transportation system for foodservice industry: modelling factors influencing customer's intention to adopt drone food delivery (DFD) services. *Sustainability, 14*, 2852.

Kamgang, S. E., Boiral, O., Guillaumie, L., & Brotherton, M.-C. (2023). Responsible sourcing in the food industry: A scoping review. In *Environment, development and sustainability* (Vol. 26, pp. 1–25).

Küfeoğlu, S. (2022). *SDG-13: Climate action* (pp. 429–451). Springer.

Leite, H. R., & Vieira, G. E. (2015). Lean philosophy and its applications in the service industry: A review of the current knowledge. *Production, 25*, 529–541.

Li, M., Jia, N., Lenzen, M., Malik, A., Wei, L., Jin, Y., & Raubenheimer, D. (2022). Global food-miles account for nearly 20% of total food-systems emissions. *Nature Food, 3*, 445–453.

Menon, A., Stojceska, V., & Tassou, S. A. (2020). A systematic review on the recent advances of the energy efficiency improvements in non-conventional food drying technologies. *Trends in Food Science & Technology, 100*, 67–76.

Mensi, A., & Udenigwe, C. C. (2021). Emerging and practical food innovations for achieving the Sustainable Development Goals (SDG) target 2.2. *Trends in Food Science & Technology, 111*, 783–789.

Otto, S., Strenger, M., Maier-Nöth, A., & Schmid, M. (2021). Food packaging and sustainability–consumer perception vs. correlated scientific facts: A review. *Journal of Cleaner Production, 298*, 126733.

Pandey, S., Sharma, K., & Gundabala, V. (2022). Antimicrobial bio-inspired active packaging materials for shelf life and safety development: A review. *Food Bioscience, 48*, 101730.

Rabadán, A., González-Moreno, Á., & Sáez-Martínez, F. J. (2019). Improving firms' performance and sustainability: the case of eco-innovation in the agri-food industry. *Sustainability, 11*, 5590.

Roy, P., Mohanty, A. K., Dick, P., & Misra, M. (2023). A review on the challenges and choices for food waste valorization: Environmental and economic impacts. *ACS Environmental Au, 3*, 58–75.

Rufí-Salís, M., Petit-Boix, A., Villalba, G., Sanjuan-Delmás, D., Parada, F., Ercilla-Montserrat, M., Arcas-Pilz, V., Muñoz-Liesa, J., Rieradevall, J., & Gabarrell, X. (2020). Recirculating water and nutrients in urban agriculture: An opportunity towards environmental sustainability and water use efficiency? *Journal of Cleaner Production, 261*, 121213.

Sardar, M. A., Khan, M. I. A., Salman, M., & Ullah, I. (2023). Farm animal welfare as a key element of sustainable food production: Animal welfare and sustainable food production. *Letters in Animal Biology, 3*, 01–08.

Sayas-Barberá, E., Pérez-Álvarez, J. A., Navarro-Rodríguez de Vera, C., Fernández-López, M., Viuda-Martos, M., & Fernández-López, J. (2022). Sustainability and gender perspective in food innovation: foods and food processing coproducts as source of macro-and micro-nutrients for woman-fortified foods. *Food, 11*, 3661.

Shou, Y., Shao, J., Lai, K.-h., Kang, M., & Park, Y. (2019). The impact of sustainability and operations orientations on sustainable supply management and the triple bottom line. *Journal of Cleaner Production, 240*, 118280.

Singh, R., Bembem, K., Solanki, C., & Dawange, S. (2023). Statistical tools and techniques for food processing data analysis: An overview. *Hypothesis, 2*, 4.

Sosa-Martínez, J. D., Montañez, J., Contreras-Esquivel, J. C., Balagurusamy, N., Gadi, S. K., & Morales-Oyervides, L. (2023). Agroindustrial and food processing residues valorization for solid-state fermentation processes: A case for optimizing the co-production of hydrolytic enzymes. *Journal of Environmental Management, 347*, 119067.

Sreedevi, P., Jayachandran, L. E., & Srinivasa, R. P. (2020). Application of high-pressure processing for extending the shelf life of sugarcane juice under refrigerated conditions. *Journal of Food Processing and Preservation, 44*, e14957.

Sunny, J., Undralla, N., & Pillai, V. M. (2020). Supply chain transparency through blockchain-based traceability: An overview with demonstration. *Computers & Industrial Engineering, 150*, 106895.

Suri, S., Singh, A., & Nema, P. K. (2021). Recent advances in valorization of citrus fruits processing waste: A way forward towards environmental sustainability. *Food Science and Biotechnology, 30*, 1–26.

Tyagi, P., Salem, K. S., Hubbe, M. A., & Pal, L. (2021). Advances in barrier coatings and film technologies for achieving sustainable packaging of food products–a review. *Trends in Food Science & Technology, 115*, 461–485.

Walker, T. R., McGuinty, E., Charlebois, S., & Music, J. (2021). Single-use plastic packaging in the Canadian food industry: Consumer behavior and perceptions. *Humanities and Social Sciences Communications, 8*, 1–11.

Willer, H., Trávníček, J., & Schlatter, S. (2024). The world of organic agriculture. Statistics and emerging trends.

Yadav, B., Pandey, A., Kumar, L. R., & Tyagi, R. D. (2020). Bioconversion of waste (water)/ residues to bioplastics-a circular bioeconomy approach. *Bioresource Technology, 298*, 122584.

Chapter 11
Quality Control and Food Safety

Muhammad Ammar Khan, Nguyen Van Hue, Xiang Wang, Sher Ali, Adil-ur-Rehman, and Hafiz Talha Hafeez

1 Introduction

1.1 Food Quality

In an integrated food and bioprocessing industry (IFBI), food quality refers to the degree to which a food product meets the standards established by the IFBI itself. These standards are developed based upon various requirements as follows:

1. Law requirements
2. International standards
3. Consumer requirements
4. Own requirements of the IFBI

M. A. Khan (✉) · Adil-ur-Rehman · H. T. Hafeez
Faculty of Agriculture & Environment, Department of Food Science & Technology, The Islamia University of Bahawalpur, Bahawalpur, Pakistan
e-mail: ammar@iub.edu.pk

N. Van Hue
University of Agriculture and Forestry, Hue University, Hue, Vietnam

X. Wang
School of Health Science and Engineering, University of Shanghai for Science and Technology, Shanghai, China

S. Ali
Faculty of Agriculture & Environment, Department of Food Science & Technology, The Islamia University of Bahawalpur, Bahawalpur, Pakistan

Institute of Food Science and Technology, Sindh Agriculture University Tando Jam, Hyderabad, Pakistan

A. S. Jatoi (ed.), *Integrated Food and Bioprocessing*,
https://doi.org/10.1007/978-3-031-87209-9_11

Traditionally, food industry chooses to develop standards related to physical, chemical, and microbiological aspects of products. Since, sustainability has been a growing concern these days, hence an IFBI focuses on all these aspects. The purpose of conforming all these criteria is to enhance product acceptability, increase IFBI's profit, and ensure product safety.

Increasing knowledgeable consumers, and their demands for health- and environment-friendly foods, the demand for quality and health-friendly foods has risen during recent years. Consumer perceptions are central to food quality, as these perceptions significantly influence their buying decisions. During the last few decades, consumers concerns regarding health-friendly products and sustainability have reshaped their food buying decisions. However, consumer interpretations of quality often contradict with actual quality.

Food quality evaluation involves two stages: pre-purchase and post-purchase. Food packaging design and labels significantly affect consumers perceptions of product quality; however, variations may exist among consumer perceptions due to noncompliance of government laws, consumer misunderstanding, and changing industry practices. Labelling helps consumers make informed choices, by reducing the information gap between producers and consumers, particularly through detailed product information, thereby increasing trust.

The measurement of food quality in an industry is quite different from measuring quality in labs of research organizations and academic institutions. The intensity of industrial testing is high towards quality attributes of product which consumer can directly assess, i.e. product packing material, printing quality, sealing ability, etc. Moreover, industry focuses on sensory attributes of product including product color, taste, aroma, and texture. However, the intensity of determining the quality parameters which are usually published in research articles is lower in industrial labs. These include determination of oxidation stability, nutritional contents, microbial number, pH, etc. Moreover, detailed quality testing is not done in most of the industry, which is related to volatile compounds related to taste and aroma, rheology, proteins quality, water holding capacity, detailed product textural attributes, etc.

1.2 Food Safety

Food safety refers to the measures taken to prevent foodborne diseases and injuries from the consumption of food. It involves managing the risks posed by biological, chemical, or physical agents, which are also called food hazards. Unsafe food, which can lead to a foodborne disease such as food poisoning, is a result of these hazards. The study of food safety encompasses all factors that can make food unsafe for consumption. Anything that is not considered as food or food substance is considered as foreign matter. For example, a broken crockery piece, bone, piece of egg shell, nail, hair, etc. in food items are considered as foreign matter. A hazard associated with accidental or intentional additional of any foreign object into the food is known as physical hazard. Detergents, sanitizers, pesticide residues, etc. in food items are considered as chemical hazards. Pathogenic or spoilage microorganisms found in food items are considered as biological hazards. Few other hazards include allergens. An allergen is a food item upon eating which the body recognizes it as a foreign substance and initiates its immune response. The problem is called allergy. In the USA, eight food items are considered allergens, while in the UK, 14 items have been listed as allergens. Allergies can cause a mild discomfort to severe life-threatening reactions. Food intolerance is a similar term to allergy; however, human body does not invoke immune response upon eating such food items. Famous examples of food intolerance include lactose intolerance in milk and dairy products, as well as gluten intolerance in wheat and bakery products. The food borne diseases can be classified as injuries, infections, intoxications, and toxico-infections. Injuries are caused by physical hazards, infections are caused by pathogenic microorganisms, and intoxication is caused by intake of toxic chemicals. Toxico-infection refers to the infection in which microorganisms produce toxin in human gut.

1.3 Sustainability

Food safety and quality are integral components of sustainable food systems; therefore, both are mandatory for an IFBI. These days there is growing debate about health- and environment-friendly foods, because consumers consider clean-label foods as health-friendly (Raposo, Ramos, Raheem, Saraiva, & Carrascosa, 2021). The concepts of clean-label foods, organic foods, or green foods are synonymous. These foods use organic ingredients and additives in place of synthetic ingredients or additives to manufacture products and to improve their technological traits. Due to this practice, it is thorough that their quality and safety enhances, that is why they are considered sustainable.

2 Food Safety Laws and Certifications

2.1 Regulatory Requirements

Compliance to law is mandatory for an IFBI. Food laws must be laid down to ensure provision of safe and wholesome food products for consumption by creating effective regulatory system for manufacturing operations, storage facilities, supply chain, sale and import of foods. The most important requirement is to implement sanitary and hygienic standards. Moreover, an IFBI should not manufacture, store, sell or distribute any unsafe or misbranded foods. Each IFBI must hold a valid license to manufacture, store, distribute, sale and import the relevant food items. The license must be renewed after regular auditing after compliance to the latest provisions of the law. Moreover, the IFBI must procure food and non-food raw materials used for manufacturing and packing of food items from licensed vendors, and ensure their safety according to the law. The IFBI must not produce, sell or sale any food or non-food items prohibited by the law. It must not employ any person with any kind of infectious disease. An IFBI must give a warrantee about the nature and quality of the food items it is manufacturing. An IFBI must ensure that all the food items it is offering for sale are tested regularly for safety according to the law.

The IFBI must employ an adequate number of certified food professionals according to its size and nature of business. An IFBI must give opportunities to emerging food professionals for food safety training in professional environment to let them get ready for professional environment. For this purpose, industry-academia linkage may be promoted. All workers involved in handling food employed in an IFBI must be trained for the requirements of food safety. The company must adapt procedures for safe disposal of wastes, and ensure judicial use of resources to ensure environmental sustainability. The record of all the customer complaints and feedbacks regarding food safety must be kept, and appropriate root cause analysis should be done for each case. Appropriate corrective and preventive actions must be taken for ensuring food safety. The IFBI must have an adequate traceability system in function, and conduct mock traceability on regular basis.

2.2 Quality Management System

Industry adopts some quality certifications to increase consumer trust in their products. One of the most famous quality certifications is ISO 9001. ISO 9001 is a quality management system, which states "Do what you write, and write what you do." The message is quite simple that the IFBI initially decides some standards according to government regulations, international standards, consumer demands, and according to its own brand requirements. The IFBI will write all these standards, and implement them in all the stages of their procurement, processing, storage, and distribution. According to the second part of this message, the IFBI monitors that all the standards are implemented at all the stages of procurement, processing, storage, and distribution, and writes the monitoring data in designated and controlled documents. The monitoring is done according to the standards decided and written in the initial stage, and monitoring frequency is also decided. The ISO 9001 systems allows IFBI to improve its standards according to changing legislation, international standards, consumer demands, and own requirements. ISO 9001 was first introduced in 1987 by the International Organization for Standardization (ISO). Initial version was named ISO 9001:1987, and subsequent versions were ISO 9001:1994, ISO 9001:2000, ISO 9001:2008, ISO 9001:2015. It is expected that the next revision of ISO 9001 will be published at the end of 2025. ISO 9001 is based on seven quality management principles, including customer focus, leadership, engagement of people, process approach, improvement, evidence-based decision making, and relationship management. It is, therefore, considered a quality management system or QMS. It is organized into several chapters, including scope, normative references, terms and definitions, context of the organization, leadership, planning, support, and operation. Each chapter contains specific requirements for establishing, implementing, maintaining, and continually improving a quality management system. Today, many food industries have been certified for QMS, and consumers trust this system.

2.3 Food Safety Management System

Ensuring food safety is a global responsibility, with governments enforcing regulations that the food industry must adhere to. Public health is the prime concern for governments. Moreover, public health problems cause considerable socioeconomic impact. IFBIs may also suffer considerable losses in terms of loss of business, expenses on recalling unsafe products, disposal, penalties, legal costs, disruptions in the food supply chain, and reputation damage due to foodborne outbreaks caused by their unsafe food (Hussain & Dawson, 2013). Hence, IFBIs take food safety as the most important issue for them. Therefore, IFBIs use the Hazard Analysis and Critical Control Point (HACCP) system, which is a famous system used for ensuring food safety in food industries. HACCP is a systematic approach for ensuring food safety in food industry is by identifying, controlling, and preventing biological, physical, and chemical hazards. The US National Aeronautics and Space Administration (NASA) introduced the HACCP system in the 1960s, in collaboration with the Pillsbury Company and the US Army Laboratories; however, food industry adopted it in 1971. Initially, the HACCP system had three principles, but now it has seven principles. These include conducting a hazard analysis, identification of critical control points (CCPs), establishing critical limits, monitoring CCPs, establishing corrective actions, implementing verification procedures, and lastly record-keeping and documentation. In this system, CCP holds the central place, which refers to a point at which control measures can be applied prevent, eliminate, or reduce a food safety hazard to an acceptable and tolerable level. In 2005, ISO developed ISO 22000 standard, which was an integration of HACCP with ISO 9001. Since, ISO was known as QMS and HACCP was known as food safety system, hence ISO 22000 was named as food safety management system (FSMS). In 2009, the Foundation for Food Safety Certification developed FSSC 22000 certification scheme. The FSSC 22000 scheme was built on the ISO 22000 standard, however, it had additional requirements. Consumers accept HACCP, ISO 22000, as well as FSSC 22000 as symbols of food safety, and food industries globally get certified for all of these.

3 Principles of Food Preservation

Food spoilage is a process in which food quality, texture or nutritional value of the food deteriorates and becomes unfit for human consumption. There are four main principles of food preservation, including control of microorganisms, autolysis, physical damage, and pests and rodents.

An IFBI must ensure that entry points of microorganisms must be strictly controlled. For this, a comprehensive sanitation program must be implemented in the IFBI. Moreover, safety of raw materials must also be monitored and appropriate correct actions must be taken. Furthermore, the food handlers suffering from any kind of contagious illness such as cold, cough, any open wound etc. must not be allowed to enter food manufacturing premises. Processing temperatures that kill pathogenic and spoilage microorganisms must be ensured during processing. After processing, cooked product must be shifted to refrigeration within first 2 h, otherwise after 4 h that product must be discarded. In addition, cross contamination must be prevented during food handling and storage. Particularly, storage time, temperature, and humidity matter. Ingredients and water used for processing must be safe and free from contamination. Segregation of cooked and uncooked foods must be done to avoid cross contamination. Moreover, traffic pattern and uniform code must be followed to avoid contact of workers of cooking and non-cooking processing areas. Air flow will be from cooking area towards non-cooking areas. Packaging material used for food items must be tested for microbial safety. Finally, recommended storage conditions must be ensured to avoid microbial growth.

Autolysis takes place due to enzymatic activity. Primarily, blanching is used to denature enzymes responsible for autolysis. Moreover, autolysis may also be controlled by controlling the abiotic conditions during storage such as temperature, humidity, etc. In addition, additives such as antioxidants can be added in food products to prevent autoxidation.

Physical damage can take place due to improper handling of food. Moreover, wrong selection of equipment and packing material may also cause physical damage to food items. Incorrect raw material selection must result in shorter shelf life. Moreover, additives such as essence, flavor etc. can spoil food if not used in right quantity. Incorrect processing method can lead to major changes in the final cooked product causing them spoil earlier. Residues of detergents and sanitizing agents can come in contact with food at any stage of processing. Low-quality packing material could also cause food spoilage. During processing, foreign matter such as nails, hair, jewelry, machine parts, etc. may drop into food items, rendering them unfit for consumption. Moreover, use of non-food grade utensils can result in decrease of shelf life and increases the risk of metal contamination of the food article. Finally, lack of following FIFO (first in first out) during storage may result in distribution and sale of products with variable expiry dates, hence increase chance of product expiry.

Final threat are pests and rodents. Food attracts pest infestation. The body fluids such as urine, fickle matter, etc. of rodents or pests can get into the food processes and make food unsafe for consumption. Moreover, improper waste disposal can lead to food spoilage or food contamination as waste produce is the basic breeding ground for flies and other pests and rodents. In addition, improper drainage system may lead to accumulation of waste product which in turn can lead to pest and rodent infestation. Therefore, an integrated pest management setup must be established in an IFBI. Three lines of defense may be installed. Within processing facility, non-chemical line of defense is installed. It includes installation of rat trap, lizard glue trap, rat cages, etc. at the entry and exit points, as well as along the waste water drains. However, non-toxic cockroach gel may be applied within electrical parts of processing equipment. The second line of defense will also be non-chemical, and along the wall of processing facility. The third line of defense will be installed along the external boundary of the processing plant. Chemical baits can be used for it.

4 Major Threats

4.1 History of Foodborne Illness

Food scarcity in the prehistoric life shaped human evolution. About, 12,000 to 10,000 years ago, humans lived like nomadic hunters and gatherers, and used to follow water reservoirs and animal migrations. However, domestication of crop plants and livestock animals proved to be reliable food source for them, and led to farming and denser populations near water reservoirs. However, close contact with animals and poor sanitation resulted in disease transmission, while travel for trade spread those diseases globally. The causes of diseases were microorganisms including bacteria, viruses, protozoans, helminths, and prions. Moreover, toxicants produced by these microorganisms also became the reason for illness of thousands of humans. Insects, rodents, and pests became the vectors for spread of these diseases.

4.2 Infections

Globally, etiology of 32% of all reported outbreaks is still unknown, viruses stand at number two with 30% outbreaks, followed by bacterial etiologies (28%).

Viruses are acellular organisms. They require a host to reproduce. Few important foodborne viruses include norovirus, hepatitis A virus (HAV), hepatitis E virus (HEV), and human rotavirus (HRV).

Bacterial are single cellular organisms, which are prokaryotes. Some bacteria are beneficial, few do not harm humans, but few are pathogenic. The can have rod, spherical, and spiral shapes, and their sizes can vary from 0.2 μm to a relatively large 750 μm. They can be classified as aerobes, obligate anaerobes, facultative, and microaerophilic. They can also be categorized as psychrophilic (−4 °C optimum), mesophilic (−39 °C optimum), and thermophilic (−60 °C optimum). Most common infectious foodborne pathogens are *E. coli*, Salmonella, shigella, yersinia, clostridium, vibrio, and coliform. *E. coli* O157:H7 is the most notorious pathogen among six classes of *E. coli*, which include enterohemorrhagic *E. coli* (EHEC), enteroinvasive *E. coli* (EIEC), enteropathogenic *E. coli* (EPEC), enterotoxigenic *E. coli* (ETEC), diffuse-adherent *E. coli* (DAEC), and atypical enteropathogenic *E. coli* (EAEC). Typhoid is caused by salmonella species. Some bacteria can also cause toxico-infection, which are mainly spore formers, and include *Clostridium perfringens*, *C. botulinum*, *Bacillus cereus*.

Protozoa are eukaryotic, single-celled organisms. They reproduce asexually or sexually. They form cysts to survive harsh environments. Their examples include Cryptosporidium spp., Giardia spp., *Cyclospora cayetanensis*, *Toxoplasma gondii*, *Entamoeba histolytica* (amoebiasis), and *Sarcocystis hominis*/ *S. suihominis*.

Fungi are mostly harmless but can produce toxins (mycotoxins) that contaminate food. Mushrooms are also fungi.

Helminths are parasitic worms. They are larger and have more complex structure than protozoa. Their examples include tapeworms (phylum Platyhelminthes), roundworms (phylum Nematoda), and flukes (phylum Annelida).

Prions are misfolded proteins that cause fatal brain diseases. Once consumed, they can corrupt other proteins by inducing protein misfolding. This can result in plague formation within critical neurologic tissues and organs. Examples of prions include bovine spongiform encephalopathy (BSE), which is famously known as Mad Cow Disease.

4.3 Toxicology

Food poisoning can happen from intoxication as well as toxicoinfection. Toxins are subcategories of toxicants produced naturally from biological sources. The famous bacterial toxins include staphylococcal enterotoxins, botulinum toxins, *Bacillus cereus* enterotoxin (cereulide), and bacterial histamines (Scombroid Fish Poisoning). The famous fungal toxins include aflatoxins (B1, B2, G1, G2), ochratoxin A, ergot alkaloids, fumonisins, trichothocenes, patulin, and yellow rice toxins. Mushrooms are also fungi. They are mostly famous for being used for the preparation of savory edible dishes; however, some mushrooms are poisonous. These include psilocybin and cyclopeptides (amanitin). Shell fish poisoning can take place because these feed on algae. Algal toxins may get deposited in these fishes. Most important algal toxins include ciguatoxins, okadaic acid, dinophysis toxins, brevetoxins, saxitoxins, and domoic acid. Moreover, industrial effluents in the lakes and water reservoirs may also result in toxin accumulation in fishes. Minamata disease is famous example of bioaccumulation of methyl mercury in fishes due to addition of industrial waste water in lakes.

5 Comprehensive Control Plan

5.1 *Layout of Processing Facility*

An IFBI facility should be well-designed to ensure safe and hygienic food production. Surroundings should be clean, well-maintained, and have proper drainage. The design should prevent food hazards, facilitate maintenance, cleaning, and disinfection, provide adequate workspace and pest control measures. Building materials should be non-absorbent, washable, smooth, light-colored. Equipment and utensils should be able to withstand food, water, heat, and cleaning procedures. These must be impervious, non-toxic, corrosion-resistant, and non-flavor transferring materials. It must be designed to achieve required food temperatures quickly. An IFBI must have appropriate facilities of storage, water supply, dishwashing, ventilation, drainage, and waste management. Moreover, there must be facilities for staff to maintain cleanliness (changing areas, locker rooms, washing stations, toilets, etc.).

5.2 *Raw Materials Inspection*

An IFBIs should be able to manage materials from source to consumption to prevent contamination and waste. Sourcing should be done from reputable suppliers who hold certificates of analysis and hygienically for their packaged ingredients and food additives. Upon receiving ingredients and raw materials at the production facility, inspection must be conducted according to the defined standards by competent professionals, and unfit materials must be rejected and returned. The passed raw materials must be immediately stored according to the type of raw material, e.g. perishables must be stored at or below 5 °C, frozen items must be stored at or below −18 °C, and dry grains at 20–25 °C. Storage areas must be controlled, safe, dry, and well-ventilated, and must be clearly designated for specific types of items only. In the storage areas, the pallet and racks should be easily accessible and their arrangement must allow delivery of older stock first, i.e. follow the system of first in first out (FIFO). The conditions of storage must be checked regularly.

5.3 *Pre-production Processing*

Food production in an IFBI may be classified as primary processing and further processing. Preproduction operations are considered preliminary operations or primary processing. Their main purpose it to get the raw materials ready for industrial processing, reduce the microbial load, remove inedible portions, and prepare spices premixes. Moreover, equipment and food touching surfaces must be sanitized to prevent cross contamination. Thawing is an important process used for frozen items prior to processing them into final recipe, for which recommended practice is to place the frozen material covered at refrigeration temperature overnight. During thawing, food quality and safety issues may arise. Water thawing should be avoided as it may significantly alter the product structure and texture. Washing, peeling, slicing, weighing, and transportation to processing area are some other relevant activities. Record of all the activities is properly maintained and kept safe for reference. Everything must be coded for any future traceability. If preliminary activities are done appropriately, they increase process safety and productivity.

5.4 Food Production

This step may be considered as further processing in an IFBI. It may also be called value-addition. An IFBI may ensure quality and safety during food production by following the standards adapted by the industry. These standards may be regarding temperature of cooking equipment, temperature of cooking oil, temperature of grill, etc. Moreover, processing times are important, as lower time than standard can result in undercooking and more time can result in overcooking. Both are detrimental for quality and safety, as undercooking may result in retention of pathogenic microorganisms in the food, while overcooking may result in loss of weight and burning of surfaces. Other important conditions may include operating pressures, which significantly impact the product structure, texture, and rheology. The standards for product temperature, size, shape, weight, and packaging material must be fulfilled to ensure conformance to already written company standards. The production is done according to consumer demand, production plans, and storage capacity. After production, the product is sent to stores under appropriate conditions.

5.5 Quality Control

In an IFBI, quality monitoring starts right from receiving of raw ingredients. Standard operating procedures are developed how to monitor the quality. One of the important methods of quality control (QC) is to use forms on which data about the conformance of production parameters like, process temperatures, product sizes, number of pieces in product packs, labelling, etc. is written. Testing is done at regular frequency and by the trained staff only. Production staff maintains all this data for each and every product in order to allow future traceability. Thermometers and weighing balances used for noting process temperatures and weights, must be well maintained and calibrated at regular intervals. Yearly third-party calibration may be performed from external sources. This data is kept for auditing. An IFBI must establish systems for internal and external audits to assess the status of food safety systems. Moreover, the quality and safety systems may be reviewed and improved according to self-evaluations, customer feedback, and updates. Since companies seek certifications for various standards from reliable third parties, requirements for each certification must be clear to the quality assurance department (QA). The third-party auditing companies cross-check and verify records of QA and QC, and evaluate the food quality and safety according to the relevant standards and systems. The certificates which these organizations grant to IFBI expire after certain time period, mostly after 2–3 years. Hence QC in an IFBI is a continuous process.

5.6 Labeling

Labelling of food products ensures traceability, informed buying decision, and consumer protection. All packaged food of an IFBI must have a label containing information about product name, nature (vegetarian/non-vegetarian), ingredients (including allergens and added water), net weight/volume, etc. In addition, in order to ensure product traceability, batch number, manufacturer information, manufacturing/packaging date, best before/use by date, and country of origin must be mentioned. Moreover, the IFBI must ensure that temperature required to storage, conditions for processing and preparation, must be mentioned on the packaging. If the product contains any notified allergen according to local law, then information about the allergen must also be written in the product packaging material. Additional information, such as claims about gluten-free status, organic/fortified claims, etc. may also be written on the packaging materials.

5.7 Storage

Safe storage of finished food products in an IFBI holds significant importance. Foods most of the time require frozen storage. The standard for frozen food products is −18 °C. The fresh foods are not immediately frozen. Initially, the temperature of the food products is brought from environmental temperature to chilling temperature, i.e. less than 16 °C. During this processing and packing of the products, this temperature is maintained in processing halls. Once the product is packed, it is transported to refrigeration rooms, where temperature is between −2 °C and 4 °C. Products stay at this temperature until their temperature drops in this zone. Subsequently, the products are sent for freezing. For freezing, some industries use individually quick-freezing facility, and such products are considered IQF products. Some industries may have blast freezers, in which food products are placed for few hours to quickly drop their temperatures. The temperature of blast freezer is usually set at −39 °C, however, temperature of the products may drop of −6 °C to −9 °C in blast freezer after few hours. Subsequently, the semi-frozen products are transferred to stores, where temperature is set between −18 °C and −24 °C. The industries where fresh products are produced, a temperature between −2 °C and 4 °C is maintained. In snacks industry, where products is cooked and inert gases are used to pack the product, room temperature, i.e. 25 °C is also suitable. Packed milk and beverages are also stored at 25 °C. Storage in egg industry varies according to the treatment. If eggs are to store for long time, temperature less than 11 °C is considered adequate. The shell coated eggs can be safely stored at room temperature for long shelf life. Besides temperature, humidity also plays important role in maintaining the safety of food products. For open food products, a humidity of 75–80% should be maintained. However, for dry products or packed products, 65–70% humidity is also adequate. Circulation of air and ventilation are necessary to remove excess heat during storage. Moreover, controlled atmospheric (CA) storage and modified atmospheric (MA) storage are two novel methods, when air composition of food stores is adjusted to selected levels. CA storage is active method, whereas MA storage is passive, i.e. mixture of selected concentrations of gases is poured in the store in the former, while in the latter, the products are covered with a membrane which allows exchange of gases with environment. The principle of both the methods is the increase of carbon dioxide gas from 0.03% to a relatively higher percent (up to 5%), but to decrease the concentration of oxygen from 21% to considerably lower level. These conditions ensure inhabitation of microbial activity due to creating non-oxidative environment. However, these conditions also prevent anaerobic oxidation due to presence of oxygen in the environment.

5.8 Transportation

Food products can get spoiled during transport. Particularly, temperature sensitive products need specific temperature conditions. The products need exactly the same temperature conditions in which they were stored in the industry, i.e. frozen products will be transported in vehicles having temperature between −18 °C and −24 °C. However, temperature fluctuations may take place in these vehicles during multiple deliveries due to repeated door opening of the vehicles. In order to ensure great traceability, computerized temperature loggers can be placed in the vehicles to monitor the temperature from time to time. Besides temperature, other requirements may include prevention of contamination from microorganism and toxic materials. For this purpose, cleaning, washing and sanitization of the vehicles should be done at a regular frequency. These chemicals should be suitable for food facility, as well as color- and odor-free. Furthermore, fresh and frozen products may not be travelled together. Finally, the vehicle must be secure from pest access.

5.9 Personal Hygiene

Maintaining personal hygiene is important for ensure safety of foods in an IFBI. Proper hygiene prevents foodborne illness. Food handlers can spread germs, so maintaining cleanliness is crucial. In this regard, IFBI may implement a personal hygiene training protocol to increase awareness and practices which ensure their own and consumers health. Workers may be trained to ensure that they keep their hair restrained by using hair nets during the processing, ensure nail trimming, avoid jewelry, and wear clean uniforms. Moreover, IFBI must ensure that all employees change clothes, as well as wash and sanitize hands and shoes before entering the processing area. There must be separations between the washing and sanitization areas from the processing areas. In addition, proper hand washing protocols must be adapted, i.e. wash hands with a regular frequency with soap and warm water. After sanitation, wear disposable working gloves and change gloves at a regular frequency. Workers must avoid touching any surfaces that can cause contamination. Workers must be trained to follow proper hand washing and sanitization protocols after using the restroom, handling raw / uncooked food, cleaning, putting on/taking off gloves, eating/drinking, touching face/body. Workers with any visible symptoms of fever, coughing, itching, skin problems, diarrhea, vomiting, etc. should not be allowed to enter the workplace, and regular medical checkups must be promoted among the food handlers. AN IFBI must establish clean restrooms and locker rooms, where workers and visitors wear protective clothing and wash hands before entering food processing areas.

5.10 Cleaning and Disinfection

An IFBI must establish a proper system of cleaning and disinfection. Standard sanitary operating procedures (SSOPs) must be established and implemented in the processing facility.

The SSOPs must clearly designate a cleaning schedule, frequencies, procedures, and equipment. Microbial sampling must also be done according to a set frequency and international safety standards must be followed. There must be separate group of people who clean the used utensils in such a way that no contamination takes place in the food. It is best to have a separate cleaning and sanitation section for utensils and baskets used for food handling. Clean water, food grade detergents, and scrubbers must be used to remove dirt, grease, and food particles from the tools and equipment in which food products are being prepared. Sanitization must also be performed after washing with detergent using chlorine or QMBA based sanitizers according to proper dilution. In an IFBI, abrasive cleaners and acids must not be used, as they can cause damage to food contact surfaces and utensils.

5.11 Waste Disposal

An IFBI must establish a proper waste disposal system. Drainage of wasted food contributes to the major portion of effluent from the food industries. These drains often throw water into mainstream lakes and canals; hence they must not have any chemicals which can damage the lake ecosystem. An IFBI must ensure the establishment of wastewater treatment protocol and water should only go out of the industry when it is safe to go out. Moreover, gratings must be installed on all the drain exits, so that no food particle may go out of the IFBI. A lot of food waste can be recycled and reused for other purposes. So, an IFBI must ensure none of its food waste contributes to environmental degradation. Waste bins must be placed throughout the processing areas, and they must be cleaned at a regular interval. Moreover, none of the waste must be thrown out of processing facility. Neither it must be disposed in open air, not it must be left out for birds, rodents, and pests. The waste bins should also be distinguished to contain biodegradable and non-biodegradable wastes. The waste bins must be installed away from the walls to avoid hiding places for insects, pests and rodents. Biodegradable platic bags must be installed inside the bins, which must be regularly changed. After a regular frequency, the waste bins must also be washed with hot water using detergent and sanitized with disinfectant.

5.12 Pest Control

Pests are considered vectors for spread of various diseases. An IFBI must implement a proper integrated pest management (IPM) system in its food production facility. It can comprise of mechanical, biological, and chemical control. A certified pest control professional must be hired in a food industry, who must train his team for proper implementation of pest control plan according to the approved IMP system of the IFBI. Only the approved pesticides for food industry must be used in an IFBI, and all the data of chemical and mechanical control of pests must be kept available for audit. The chemical store must be away from the processing facility, and arrival, storage, and usage of all the chemicals must be according to approved protocols. No pesticide should be used inside the food processing facility. Cockroaches, rodents, lizard, flies, and ants are primary pests which are tend to be prevalent in food processing facilities. In addition, there is a huge class of stored gain pests as well, which can be hazardous in food processing from those grains used as ingredients. A 4-pronged approach can be used for pest control, i.e. deny entry, deny shelter, deny food, and destroy pests. Since cockroaches prefer warm and moist areas, they are mostly present in the cooking areas. The preventive measure for them is to seal all the cracks in the walls, clean almirahs, regularly check all the electrical panels of machines. Cockroach gels are safe to use in processing areas, so they can be used in the electrical panels of machines if cockroaches hide in them. Rodents enter the processing facilities along the walls, so mechanical traps must be placed along the walls of processing facilities, both inside and outside. Rodents may also enter through the drains, so drain gratings must be installed on the exit points of the drains. Drains must be regularly cleaned, as food particles in the drains attract the rodents. Rodents dig their burrows in the compound, so all such burrows must be eradicated. In an IFBI, three lines of defense may be installed, a non-chemical line of defense is comprising mechanical traps and glue baits, inside processing facility, the second line of defense containing non-chemical traps along the wall of processing facility, and the third line of defense containing chemical baits along the external boundary of the processing plant. The chemical line of defense along the external wall of the factory can avoid the entry of new rodents from external surroundings field and areas. Lizards stay on the electrical cable trays, so regular cleaning of electrical cable trays must be performed. In addition, lizards can be controlled using glue baits, which attract lizards and trap them. These can be then thrown away from the processing areas. Flies breed in filth, hence an IFBI must ensure that all the waste bins remain covered near the processing facility and all the waste is dumped in an environment-friendly manner. Prevention of food particles in drains also saves the harboring of flies on the drains. In addition, chemical spray must be done in the drains outside the processing facility to avoid flies and mosquito rearing and growth. Ants are also one of the most important vectors of foodborne diseases, so their entry and breeding sites must be controlled.

5.13 Training

An IFBI needs to establish a proper training mechanism for food handlers to reduce foodborne illnesses. The foremost important is personal hygiene of worker who are doing the food processing or handling the processed foods directly. An ill person in the processing hall can contaminate tons of product in the processing hall, which could result in illness threat to thousands of consumers. Moreover, all the workers must be trained how to handle food in storage and distribution. Furthermore, periodic trainings of senior management should be conducted regarding the food safety regulations. Results of all the trainings should be evaluated or effectiveness on regular interval, and all the training records should be documented and remain available for audit.

6 Emerging Food Processing Methods

Historically thermal processing and chemical additives were used in preservation of food products. However, novel processing methods are continuously substituting traditional processing methods, mainly due to greater food quality and safety control. Particularly, high-pressure processing (HPP) destroys detrimental microbes and indigenous enzymes by applying high hydrostatic pressure (100 MPa to 1000 MPa) to food material consistently and simultaneously from all directions (Martínez-Monteagudo & Balasubramaniam, 2016). Pulsed electric field uses short bursts of electricity for microbial inactivation in foods (Wang et al., 2018). During microwave processing, dielectric heating produces electromagnetic waves. When food is exposed to a microwave field, the water molecules in the food absorb the microwave energy and convert it into heat that kills microorganisms (Laguerre & Hamoud-Agha, 2022). Ohmic heating involves the passage of an alternating current through food, that generates heat in the product due to molecular friction, causes oscillation of molecules and ions, thus kills microorganisms (Kaur & Singh, 2016). During cold plasma technique involves treatment with plasma that contains high-energy ions and electrons which upon exposure to foods produce reactive species that damage the cell membranes of microorganisms (Rathod et al., 2021). Hurdle technology involves combination of two or more technologies for effective and sustainable microbial control. Other novel technologies include radio frequency heating, ultrasonic processing, active packaging, etc.

7 Conclusion

In an integrated food bioprocessing industry (IFBI), consumer trust can be achieved by controlling food quality, safety, and sustainability. For this, an IFBI must follow the governmental regulations and adapt international certifications, including quality management system (QMS) and food safety management system (FSMS) throughout its processing facility and supply chain. Standard operating procedures (SOPs) for the control of microorganisms, enzymatic activity, pests, and physical damage need to be established, and the IFBI must ensure provision of essential resources for successful implementation of those SOPs. Moreover, effective quality assurance (QA), quality control (QC) system must be established for standardization, monitoring, and verification of all the standards of processing and storage. Furthermore, all the stakeholders must be trained for food safety, quality control, and personal hygiene. The IFBI must prefer novel sustainable processing methods over traditional methods to ensure higher quality control, food safety and sustainability.

References

Hussain, M. A., & Dawson, C. O. (2013). Economic impact of food safety outbreaks on food businesses. *Food, 2*(4), 585–589.

Kaur, N., & Singh, A. (2016). Ohmic heating: Concept and applications—A review. *Critical Reviews in Food Science and Nutrition, 56*(14), 2338–2351.

Laguerre, J. C., & Hamoud-Agha, M. M. (2022). Microwave food processing: Principles and applications. In *Thermal food engineering operations* (pp. 301–347).

Martínez-Monteagudo, S. I., & Balasubramaniam, V. (2016). Fundamentals and applications of high-pressure processing technology. In *High pressure processing of food: Principles, technology and applications* (pp. 3–17).

Raposo, A., Ramos, F., Raheem, D., Saraiva, A., & Carrascosa, C. (2021). *Food safety, security, sustainability and nutrition as priority objectives of the food sector* (Vol. 18, p. 8073). MDPI.

Rathod, N. B., Ranveer, R. C., Bhagwat, P. K., Ozogul, F., Benjakul, S., Pillai, S., & Annapure, U. S. (2021). Cold plasma for the preservation of aquatic food products: An overview. *Comprehensive Reviews in Food Science and Food Safety, 20*(5), 4407–4425.

Wang, Q., Li, Y., Sun, D.-W., & Zhu, Z. (2018). Enhancing food processing by pulsed and high voltage electric fields: Principles and applications. *Critical Reviews in Food Science and Nutrition, 58*(13), 2285–2298.

Part V
Innovation and Future Directions

Chapter 12
New Product Development in IFB

Abdul Sattar Jatoi, Arshad Iqbal, and Bazil Owais Nasir

1 Introduction

New product development (NPD), a multidisciplinary process in integrated food and bioprocessing, combines concepts from food science, biotechnology, and engineering to create innovative food products. In addition to being safe and nutritional, this process involves developing new food products that meet consumer aspirations for sustainability, convenience, and unique experiences. By using bioprocessing methods including fermentation, enzymatic reactions, and microbial systems, the development process can increase food quality, prolong shelf life, and add new useful components.

Basic materials can be transformed into high-value food products through chemical and biological processes when bioprocessing is employed in food production. Through the use of cutting-edge technologies like bioengineering, nanotechnology, and biotechnology, this integration promotes the manufacture of healthier, more sustainable food items with enhanced nutritional profiles and sensory attributes. Furthermore, via the use of developments in waste management and the circular economy, food processing by-products can be converted into beneficial resources like biofuels, bio-based packaging, or other bioproducts.

New product development in integrated food and bioprocessing is greatly impacted by shifting consumer trends and regulatory environments. There is a growing demand for plant-based, clean-label, and functional foods, and sophisticated methods are needed to preserve the desired sensory and health benefits. Regulatory issues ensure environmental sustainability, food safety, and quality control throughout the NPD process. Combining food and bioprocessing results in a

A. S. Jatoi (✉) · A. Iqbal · B. O. Nasir
Department of Chemical Engineering, Dawood University of Engineering and Technology, Karachi, Pakistan
e-mail: abdul.sattar@duet.edu.pk

A. S. Jatoi (ed.), *Integrated Food and Bioprocessing*,
https://doi.org/10.1007/978-3-031-87209-9_12

Fig. 12.1 The factors of success for new product development: an overview. (Ernst, 2002)

range of creative, sustainable, and diverse food solutions that meet consumer needs and global sustainability concerns (Fig. 12.1).

2 The Role of Integrated Food and Bioprocessing in NPD

New Product Development (NPD) relies heavily on integrated food and bioprocessing because it enables the application of biological, chemical, and technical ideas to transform raw materials into innovative, high-value food products. This approach makes it possible to produce creative meals that meet consumer demands for better nutrition, sustainability, and sensory experiences. By using techniques including fermentation, enzymatic treatments, and microbial processing, integrated bioprocessing can improve food texture, flavor, and nutritional value, add functional ingredients, and ensure safety and stability.

One of the key advantages of integrated bioprocessing in NPD is its potential to increase resource efficiency and sustainability. Waste and by-products from the food manufacturing industry can be converted into valuable resources like biofuels, biodegradable packaging, or bioactive ingredients that can be added to food by using bioprocessing technology. This reduces food waste and encourages the circular economy, which aligns with the growing focus on sustainable food systems. Additionally, integrated bioprocessing can lessen the environmental effect of food production by using less energy and water thanks to improved biotechnological processes.

Moreover, integrated food and bioprocessing can be used to create new product categories that cater to market preferences, such as plant-based, functional, and clean-label meals. Thanks to biotechnological developments, food makers can create products with enhanced health benefits, such probiotics, prebiotics, and bioactive peptides, while maintaining natural and clean ingredient labeling. By ensuring food safety, extending shelf life, and enhancing flavor and texture, integrated bioprocessing contributes to the production of marketable, superior products that meet consumer preferences and regulatory standards.

3 Stages of New Product Development

The food and bioprocessing sectors employ New Product Development (NPD), a methodical set of procedures, to get a new product from concept to market. These stages ensure that the product is not just innovative but also meets consumer needs, legal requirements, and technological requirements for successful commercialization. The key stages are:

1. Idea Generation and Screening

   Potential product concepts are collected and discussed during this initial stage. Ideas may come from a variety of sources, including competitive analysis, new trends, market research, consumer feedback, and technological developments. The goal is to identify market gaps or possibilities that support business goals and consumer demand. After the creation of ideas, a screening process narrows the list to the most promising concepts by evaluating them based on technical specifications, market potential, feasibility, and alignment with corporate goals.
2. Concept Development and Testing

   After idea screening, the next step is to develop thorough conceptions for the selected ideas. Models, prototypes, or product descriptions are created in order to ascertain whether the concept is appealing to the target market. In the food and bioprocessing industries, this can mean developing a sample formulation, testing with different combinations of constituents, and assessing sensory attributes including taste, texture, and appearance. Concept testing often involves gathering feedback from focus groups, expert panels, and potential customers to refine the product idea and ensure market viability.
3. Business Analysis and Feasibility

   Before moving further, the operational and financial feasibility of the new product must be thoroughly assessed. This includes estimating manufacturing costs, pricing strategies, potential profits, and required investments. The investigation also looks at sustainability, supply chain logistics, scalability, and regulatory compliance. If the product is deemed financially and operationally viable, it advances to the next stage.
4. Product Development

   This stage involves transforming the concept into a completed product. Food scientists, bioprocess engineers, and technologists improve formulations, streamline processes, and conduct tests to create a product that meets the necessary criteria. Examples of integrated food and bioprocessing include obtaining ingredients, optimizing fermentation processes, and increasing production from laboratory to pilot plant levels. At this stage, labeling, packaging, and shelf-life stability are also discussed.
5. Testing and Validation

   A comprehensive testing procedure is carried out following development to ensure the product meets safety requirements, legal requirements, and quality standards. This includes sensory evaluations, shelf-life research, microbiological

Fig. 12.2 Strategic management of IFB. (Vega-Rosado, 2022)

testing, and, in some cases, clinical trials (for functional or health-promoting items). To verify the product's appeal, consumer acceptance tests are also conducted. Any issues discovered in this stage are resolved prior to full-scale manufacturing.

6. Commercialization

   Once the product passes all tests and satisfies all legal and market requirements, it moves into the commercialization stage. Increased production, completed packaging and distribution strategies, marketing campaigns, and product distribution to retail locations or direct-to-consumer channels are all necessary. Following launch, the product's success in the market is regularly monitored to make any necessary adjustments in response to consumer input or market developments.

7. Post-Launch Evaluation

   After launch, continuous monitoring and assessment are necessary to gauge the product's performance in terms of sales, customer satisfaction, and market share. Feedback from stakeholders and customers helps identify areas that need improvement, and supply chain or operational issues are fixed. Based on this evaluation, companies may modify, expand, or improve their product to remain competitive in the market (Fig. 12.2).

4 Innovative Techniques in NPD

Modern New Product Development (NPD) techniques are currently used in the food and bioprocessing industries to create, design, and introduce new products. These techniques improve product quality, sustainability, and consumer appeal while also increasing efficiency throughout the development process. Here are several significant state-of-the-art NPD methods:

1. 3D Food Printing

   3D food printing is a new technique that allows the construction of intricate, customized food creations by stacking materials into the proper shapes and structures. This technique allows for a considerable lot of customization, allowing for the development of unique textures, shapes, and even flavors. It is particularly useful for developing specific diets, such as gluten-free, high-protein, or vegan meals, and can be applied in domains like functional foods, where precise nutrient distribution is essential. Additionally, 3D printing may reduce food waste through innovative use of by-products or replacement parts.

2. High-Pressure Processing (HPP)

   High-pressure processing is a non-thermal food preservation technique that uses high pressure to deactivate harmful bacteria, pathogens, and enzymes without compromising the food's flavor or nutritional value. This method extends the shelf life of food products without compromising their freshness, flavor, or texture. HPP is particularly important in NPD for clean-label products since it removes the need for chemical preservatives, meeting consumer demands for less processed and more natural foods.

3. *Fermentation and Bioprocessing Innovations

   Innovations in fermentation technology have revolutionized NPD by enabling the production of new functional foods, plant-based proteins, and probiotics. Precision fermentation can be used to create highly specialized components with enhanced nutritional or sensory properties, such as enzymes, bioactive compounds, and alternative proteins. Microbial fermentation helps replicate the sensory attributes of traditional animal-based goods by enhancing the taste, texture, and digestion of plant-based products in the growing market for plant-based alternatives.

4. Artificial Intelligence (AI) and Machine Learning

   NPD is using AI and machine learning to improve formulas, predict consumer preferences, and speed up product design. AI-driven systems can analyze large statistics on market trends, ingredient performance, and consumer input, helping food makers create products that adapt to shifting consumer demands. Additionally, machine learning algorithms can reduce trial and error and speed up research and development by modeling and simulating the interactions of different chemicals.

5. Encapsulation and Controlled Release Technologies

   Encapsulation technology makes it possible to deliver and safeguard sensitive ingredients, such as vitamins, probiotics, and flavoring compounds, inside food products. This method allows for the regulated release of these compounds, ensuring their stability and bioavailability until they are consumed. Encapsulation is frequently used in NPD for beverages, fortified goods, and functional foods when preserving the stability and efficacy of ingredients is essential. By permitting the flavors to be released gradually throughout intake, it also improves sensory experiences (Fig. 12.3).

6. Sustainable Packaging and Edible Films

   Environmentally friendly packaging choices are becoming increasingly important to NPD. Examples of innovations include packaging materials that are recyclable, biodegradable, and compostable and that come from bio-based sources such as cellulose, algae, or starch. Additionally, edible films and coatings based on natural polymers are being developed to extend food shelf life while

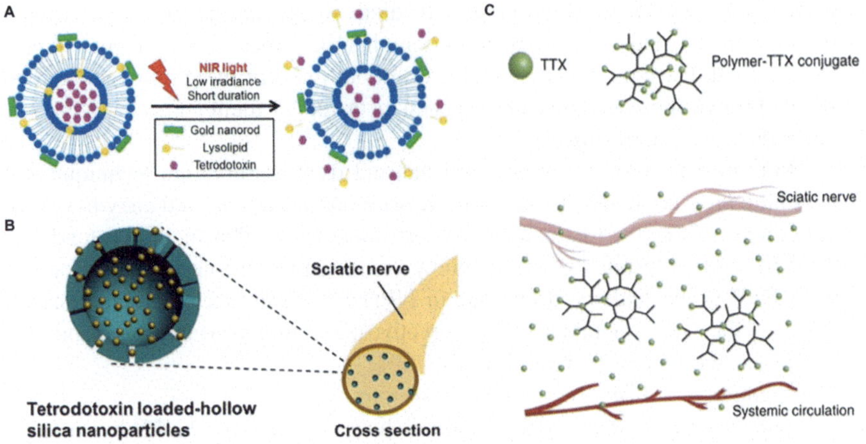

Fig. 12.3 Strategies to obtain encapsulation and controlled release of small hydrophilic molecules. (Li et al. 2020)

reducing the need for traditional packaging. In addition to satisfying consumer aspirations for sustainability, these advancements aid in reducing food waste.

7. Plant-Based and Cellular Agriculture

   The advent of plant-based and cellular agriculture has been one of the most significant advancements in NPD, allowing for the production of alternative proteins with the flavor and consistency of meat, dairy, and seafood. While plant-based technologies focus on optimizing component combinations to mimic animal-derived products using plant proteins, cellular agriculture involves growing animal cells to produce lab-grown meat. These advancements offer sustainable and moral alternatives to traditional cattle rearing, which is driving the growth of the alternative protein sector.

8. Blockchain for Supply Chain Transparency

   NPD is using blockchain technology to increase the traceability and transparency of the food supply chain. By tracking each step of the production process, from the acquisition of raw materials to the final delivery, blockchain supports claims about food safety, authenticity, and sustainability. This is particularly important for products with high consumer demand for transparency, such as fair trade, organic, or sustainably sourced food. Blockchain technology can also be used to track and verify the quality of ingredients, which benefits the end user.

These innovative techniques are transforming food production by boosting sustainability, productivity, and customer responsiveness. They enable the production of high-quality, customized food products that not only satisfy palates but also adhere to modern trends including sustainable environmental practices, ethical sourcing, and health concerns.

5 Market Trends and Consumer Preferences

Consumer tastes and market trends in the food sector are always shifting due to changes in lifestyle, environmental awareness, health consciousness, and technology advancements. These trends are crucial for guiding New Product Development (NPD) efforts to make sure that new products meet market expectations and take shifting customer values into consideration. Some of the major trends affecting the food and bioprocessing industries at the time include the following:

1. Health and Wellness Focus

Consumers' increasing focus on health and wellness when choosing foods is driving demand for products that offer nutritional benefits, improve wellbeing, and support specific health objectives. A component of this trend is the growing interest in functional foods—items that offer health benefits beyond basic sustenance, such as those enriched with vitamins, minerals, probiotics, or antioxidants. Foods that support digestive, mental, and immunological health are particularly popular as consumers seek to enhance their overall health and resilience.

Additionally, the rise in plant-based diets and the demand for clean-label products indicate customers' desire for natural, minimally processed meals free of artificial additives, preservatives, and genetically engineered ingredients. As consumers place greater emphasis on managing chronic health conditions like diabetes and heart disease, low-fat, low-sugar, and gluten-free substitutes are also growing more popular.

2. Sustainability and Ethical Sourcing

Sustainability is having a bigger impact on consumer choices as more people seek out products with less of an impact on the environment. This involves giving preference to foods that are produced using sustainable methods, use less water and energy, and generate less waste. Environmentally conscious consumers also look for companies that use fair trade and ethical sourcing methods, as well as goods with recyclable or biodegradable packaging.

Growing awareness of climate change has increased demand for plant-based foods because they generally have a smaller environmental impact than animal-based ones. Additionally, companies that promote supply chain transparency attract customers by ensuring that their products are manufactured ethically and environmentally.

3. Rise of Plant-Based and Alternative Proteins

Plant-based diets are becoming more popular because to concerns about the environment, ethics, and health benefits. The demand for plant-based alternatives to meat, dairy, and seafood has increased dramatically, and meat eaters now find these products more appetizing because to new flavor and texture advancements. Alongside vegetarians and vegans, a growing consumer group is flexitarians, or those who occasionally eat meat but prefer plant-based foods.

In response to this trend, plant-based protein sources such as pea protein, chickpeas, lentils, and algae are evolving. Cellular agriculture, or growing meat from animal cells, is gaining popularity as an alternative to conventional animal husbandry because it offers a sustainable way to produce meat without the ethical conundrums or environmental harm associated with livestock production.

4. Convenience and Ready-to-Eat Foods

As lifestyles get busier, people are increasingly choosing quick meal options that don't sacrifice taste or nutrition. Ready-to-eat (RTE) meals, meal kits, and on-the-go snacks have become more and more popular, particularly with younger consumers and working professionals. These products offer easy and quick dinner options that complement the busy lifestyle of today.

However, ease no longer comes at the expense of quality. Customers are searching for items that are fresh, convenient, and healthy, made with natural, clean ingredients. The move toward minimally processed, nutrient-dense products has led to the development of gourmet RTE meals and snacks that meet a variety of dietary needs and tastes.

5. Personalized Nutrition

The emerging concept known as personalized nutrition focuses on tailoring food products to each person's genetic composition, medical issues, and nutritional needs. Due to advancements in digital health tools and nutrigenomics—the study of how food interacts with genes—businesses are increasingly offering customized food solutions. In order to optimize their health based on their unique biological makeup and lifestyle choices, consumers are utilizing functional foods, supplements, and personalized meal plans.

Customers are driven to products that support certain health objectives, such as improved cognitive function, improved athletic performance, or weight control, and this tendency is particularly apparent in the health and wellness industry. Technology, such as wearable health devices and smartphone apps that examine biometrics, exercise levels, and food intake, is making personalized nutrition possible.

6. Transparency and Clean Labels

As consumers learn more about the ingredients in their food, they expect businesses to be more transparent about the origin, manufacturing process, and nutritional content of their goods. Clean labels, which highlight simple, natural components, have become a key marketing feature in many places. Consumers are actively seeking products with short ingredient lists that are free of artificial chemicals, preservatives, and additives.

Transparency in food production is important because consumers want to know where their food originates from, how it was grown or raised, and whether it conforms with environmental and ethical standards. By providing thorough information about their supply chains, ingredient sourcing, and production processes, businesses may boost consumer trust and loyalty.

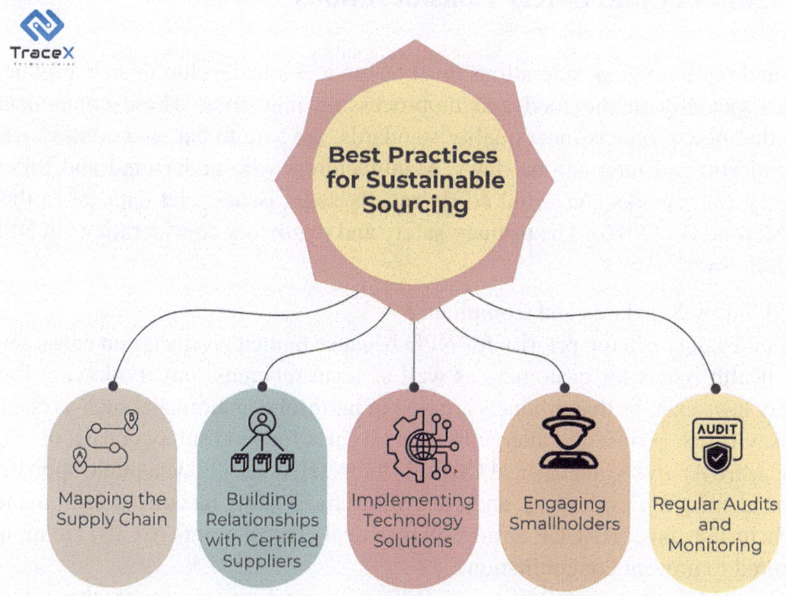

Fig. 12.4 Sustainable sourcing practices for EUDR. (Sorokin)

7. Global Flavors and Culinary Exploration

Demand for diverse cuisines and unique culinary experiences is rising as consumers grow more adventurous with their food selections. The consumption of international cuisines and fusion dishes is increasing as a result of the growing appeal of strong, exotic flavors from Asia, Latin America, Africa, and the Middle East. This trend reflects the increasing globalization of the food industry as well as consumers' desire to learn about new cuisines and cultural experiences through food.

NPD emphasizes the use of a range of authentic ingredients and production techniques in product compositions in order to meet the increasing demand for flavors from around the globe. To create products with distinctive, delectable flavors, creative applications of spices, herbs, and ingredients like tahini, miso, harissa, and turmeric are being developed (Fig. 12.4).

These trends show how customer tastes are constantly evolving and how important it is for NPD to develop to meet the shifting demands for sustainability, convenience, health, and gourmet diversity. By monitoring these trends, businesses can develop goods that not only satisfy current market demands but also anticipate future shifts in consumer behavior.

6 Regulatory and Safety Considerations

Safety and regulatory considerations must be made while developing new products (NPD), especially in the food and bioprocessing industries. These components ensure that new products meet quality standards, are safe to eat, and comply with both domestic and international laws. Manufacturers who understand and follow regulatory frameworks can avoid costly recalls, legal issues, and damage to their brand Marone PA (2016). The primary safety and regulatory considerations in NPD are as follows::

1. Food Safety Standards and Compliance

   Food safety is a top priority for NPD because tainted products can cause serious health issues for customers as well as legal repercussions. Following food safety laws ensures that products are free of harmful contaminants such as chemicals, viruses, germs, and allergens. Crucial rules for food safety consist of:

   – Hazard Analysis and Critical Control Points (HACCP): a systematic approach to identifying, evaluating, and controlling food safety hazards at each manufacturing stage. HACCP ensures that critical control points (CCPs) are monitored to prevent contamination.
   – Good Manufacturing Practices (GMPs): To produce safe foods, these basic environmental and operational elements—such as hygienic practices, cleanliness, and the use of the right equipment—are required.
   – ISO 22000: a worldwide standard for food safety management systems that outlines methods for identifying and reducing food safety hazards across the supply chain.

Following these rules helps companies stay in compliance with the law, protect the public's health, and win over customers.

2. Regulatory Approval for Ingredients and Additives

   New ingredients, additives, or bioactive compounds used in food and bioprocessing NPD must be approved by regulators to ensure their safety for consumption. The Food and Drug Administration (FDA) in the US and the European Food Safety Authority (EFSA) in the EU are two regulatory bodies in different countries that assess and approve the use of specific ingredientsPressman et al. (2017),

   – Food Standards Australia New Zealand (FSANZ).

For example, every new food ingredient, genetically modified organism (GMO), or additive that is not yet on the market must pass a rigorous approval process before being utilized in commercial food goods. Companies must give information on the ingredient's safety, toxicity, allergenicity, and nutritional impact.

3. Labeling Requirements and Claims

   Proper labeling is crucial for both regulatory compliance and providing customers with clear and accurate information. Food labeling must adhere to strict guidelines set by regulatory agencies, which include:

A food's label must clearly list all of its nutritional components, including calories, fats, carbs, vitamins, and minerals. This helps clients make educated choices regarding their nutritional needs.

In the ingredient list, each item must be listed in decreasing order of quantity. For some markets, allergens such as dairy, gluten, and nuts must be clearly marked on the label.

Claims pertaining to health and nutrition, including "rich in antioxidants" and "low in fat," are closely monitored by regulatory agencies. Claims must be scientifically verified before they may be displayed on packaging.

Sanitized Labeling: The trend toward clear labels has created.

4. Allergen Management

   Taking care of allergies is one of the most crucial safety considerations in NPD. Cross-contamination with allergens (such peanuts, shellfish, or gluten) can be quite risky for patrons who have dietary restrictions. Regulatory agencies require manufacturers to follow strict allergy management protocols throughout production and to clearly identify allergens on food labels in order to avoid contamination.

NPD processes must include allergen testing, proper equipment cleaning, and areas designated for handling allergenic materials. Food goods branded "gluten-free" or "nut-free," which are free of certain allergens, must be manufactured under stringent control to prevent accidental contamination.

5. Shelf Life and Stability Testing

   To ensure product safety and quality throughout time, extensive stability and shelf-life testing is required. Regulatory agencies require that food products have correct "best by" or expiration dates on their labels. These dates show how long the food maintains its intended level of safety, quality, and nutritional value.

Stability testing involves evaluating the product's behavior across a range of storage conditions, including changes in temperature, humidity, and light exposure. Preventing physical changes (such color or texture), chemical deterioration, or microbial development before to the designated expiration date is the goal. To safely extend the shelf life of items that fail stability testing, further preservation techniques or reformulation can be needed.

6. Novel Food Regulations

   Manufacturers who produce unique or uncommon food products, such fermented meals, cultured meats, or plant-based meat replacements, can be subject to specific innovative food restrictions. These guidelines are typically applicable to foods that have never been consumed in big quantities or that are prepared using innovative methods. Regulatory bodies like the FDA and EFSA require comprehensive safety evaluations, including toxicological testing, before approving these items for sale.

Biotechnology products, such as those made by genetic modification or precise fermentation, are guaranteed to be safe for human consumption and not to endanger the environment or public health thanks to the innovative food approval process.

Biotechnology products, such as those made by genetic modification or precise fermentation, are guaranteed to be safe for human consumption and not to endanger the environment or public health thanks to the innovative food approval process.

7. Environmental and Sustainability Regulations

   As sustainability becomes a major priority in the food industry, many regulatory frameworks now include environmental standards that must be met during product development and manufacturing. These could include guidelines for:

   Reduction of Waste Food companies are required to minimize waste and use energy, water, and raw materials efficiently. Sustainable Sourcing: Products labeled as "organic" or "sustainably sourced" must meet specific certification requirements, such as USDA Organic, Fair Trade, or Rainforest Alliance certification. Packaging Regulations: Regulators are calling for more environmentally friendly packaging solutions, like recyclable or biodegradable materials, in an attempt to reduce their impact on the environment. Food safety laws must also be followed by packaging to ensure that it protects the product from infection.

8. Traceability and Supply Chain Transparency

   For food safety and regulatory compliance, traceability across the supply chain is essential. Regulators demand that businesses maintain detailed records of the origins of raw materials, how they are processed, and where finished goods are distributed. Businesses can quickly identify the root cause of a food safety problem and take corrective action in the event of contamination or product recalls thanks to traceability.

Blockchain technology is increasingly being used to improve supply chain transparency by enabling real-time ingredient tracking and ensuring that products meet legal and ethical standards from farm to table.

9. International Regulatory Compliance

   Understanding and adhering to international regulations is essential for companies aiming to enter overseas markets. Laws pertaining to food safety, labeling, and ingredient authorization vary by country. It is not necessary to obey the regulations of one country, like as the FDA in the US, in order to comply with those of another, such as the EFSA in the EU. Companies have to handle these regulatory differences and ensure that their products meet the safety and labeling standards of each market.

By considering these safety and regulatory concerns, businesses can successfully design and promote quality, safe, and legal food items. This not only protects the public's health but also builds brand awareness and consumer trust in an increasingly regulated and competitive industry.

7 Challenges in New Product Development

New product development (NPD) in the food and bioprocessing sectors faces several challenges that could prevent new products from being introduced and adopted successfully. One of the biggest challenges is innovation under time and cost constraints. When developing a new product, extensive research, development, and testing are necessary, and these procedures can be costly and time-consuming. To be competitive, businesses need to innovate quickly, but doing so too soon could lead to poor goods, incomplete testing, or inadequate market research. Finding a balance between rapid invention and thorough development can be difficult, particularly in industries with strict laws and concerns about customer safety.

Another major issue is adjusting to shifting market trends and consumer preferences. Consumer tastes and eating habits are always evolving due to trends including sustainability, ethical sourcing, and health and wellness. In order to meet consumer requests for plant-based products, more readable labels, and personalized nutrition, businesses must continuously adapt their NPD processes. However, it's challenging to predict future trends and stay one step ahead of competitors. Companies risk investing in products that may be out-of-date or unpopular with their target markets by the time they are released. To manage this ambiguity, ongoing market research, consumer input, and flexible product development methods are required.

Last but not least, safety standards and regulatory compliance provide significant challenges for NPD. The food industry is highly regulated, and in many areas, products must meet strict safety, ingredient, and labeling standards. Complying with regulations, such obtaining approval for new ingredients or ensuring that allergen labels are correct, may be very time-consuming and costly. International product launches are more challenging for multinational corporations because they must contend with disparate laws in several jurisdictions. To maintain product innovation while ensuring compliance, product developers, regulatory experts, and supply chain partners must collaborate. This is a continuous challenge that requires dedicated resources, expertise, and teamwork.

8 Case Studies of Successful NPD in Integrated Food and Bioprocessing

Case studies of successful New Product Development (NPD) in integrated food and bioprocessing show innovative approaches that have produced both consumer acceptance and financial success. One such instance is Impossible Foods, who developed its plant-based meat alternative, the Impossible Burger, using a patented process that incorporates fermentation technology. The startup uses genetically modified yeast to manufacture heme, an iron-containing molecule found in animal muscle, which gives its burger a meaty flavor and aroma. Through extensive consumer testing and clever partnerships with restaurants, Impossible Foods was able to successfully debut its product and draw in customers looking for delicious and sustainable meat alternatives. This illustration demonstrates how innovative bioprocessing techniques can create products that appeal to customers who are concerned about the environment and their health.

Another example is Oatly, a Swedish company that specializes in dairy alternatives made from oats. The secret to Oatly's success is its unique production process, which breaks down oats into a creamy texture while preserving their nutritional value through enzymatic hydrolysis. The business was able to establish a strong presence in the rapidly growing plant-based industry by emphasizing sustainability, health advantages, and transparency in its branding. Oatly capitalized on consumer trends toward veganism and lactose-free diets to align their product offers with market demands. By using sustainable sourcing and promoting its eco-friendly manufacturing methods, Oatly has become a market leader in the dairy substitute industry, proving that sustainability can be included into NPD.

A third example is Beyond Meat, which has successfully created a range of plant-based products that mimic the taste and texture of meat by combining brown rice, pea protein, and other natural ingredients. Beyond Meat has adopted a creative strategy to cater to both vegetarians and flexitarians by employing food science and technology to produce goods that closely resemble animal-based alternatives. Their strategic partnerships with major fast-food chains, supermarkets, and food service providers increased their visibility and accessibility, which in turn led to a significant increase in client popularity. The success of Beyond Meat demonstrates how innovative food products and a strong marketing strategy can effectively meet shifting consumer expectations for healthier, sustainable protein sources.

9 Future Directions in NPD

Future New Product Development (NPD) in integrated food and bioprocessing will be heavily influenced by consumer preferences, technological advancements, and sustainability considerations. One of the key themes is the increasing use of digital technologies and data analytics in the NPD process. Big data and artificial intelligence are being used by businesses to gain a deeper understanding of consumer trends, tastes, and behaviors. Better decisions, faster product iterations, and more specialized marketing efforts are all made possible by this. Furthermore, innovations like robotics and 3D food printing are poised to revolutionize industrial processes by enabling companies to create customized goods more effectively while reducing waste and improving quality.

Another direction for the future is NPD's emphasis on sustainability and circular economy principles. Because consumers are becoming more conscious of the environmental impact of the foods they buy, businesses are developing products that prioritize eco-friendly ingredients, sustainable packaging, and waste reduction. Innovations in bioprocessing, such using insects and lab-grown meat as alternative protein sources or recycling food waste to make new products, are gaining popularity. Businesses must use life-cycle assessment techniques to examine the environmental impact of their products from farm to table in order to guarantee that sustainability is integrated into every stage of the NPD process.

Finally, the future development of NPD is expected to be significantly impacted by the rise of personalized nutrition. Advances in microbiome and genetics research are making it easier to comprehend how people differ in their nutritional needs and preferences. As customers search for products that are tailored to their needs, businesses will need to develop customizable options that satisfy various dietary requirements and health goals. This trend may lead to innovations in functional meals, meal kits, and supplements that support personalized health outcomes. Businesses that prioritize individual health and wellness in their product development strategy can increase consumer satisfaction and engagement in a market that is growing more and more competitive.

References

Ernst, H. (2002). Success factors of new product development: A review of the empirical literature. *International Journal of Management Reviews, 4*(1), 1–40.

Li, Q., Li, X., & Zhao, C. (2020). Strategies to obtain encapsulation and controlled release of small hydrophilic molecules. *Frontiers in Bioengineering and Biotechnology, 8*, 437.

Marone, P.A. (2016). Food safety and regulatory concerns. InInsects as sustainable food ingredients (pp.203-221). Academic Press.

Pressman, P., Clemens, R., Hayes, W., Reddy, C. (2017). Food additive safety: A review of toxicologic and regulatory issues. Toxicology Research and application, 1:2397847317723572.

Sorokin, S. *How are EUDR requirements implemented in companies' supply networks?.*

Vega-Rosado, L. L. (2022). The 3D IFB SWOT analysis as a strategic tool to develop entrepreneurial plans for family businesses. In *Research anthology on strategies for maintaining successful family firms* (pp. 268–294). IGI Global.

Chapter 13
Traceability and Supply Chain Management

Abdul Sattar Jatoi, Arshad Iqbal, and Bazil Owais Nasir

1 Introduction

Traceability and supply chain management are now essential elements in today's intricate food supply chain to guarantee food safety, quality, and transparency. The capacity to follow and keep an eye on food products from farm to table is known as traceability. This ability is essential for managing food recalls, identifying potential sources of contamination, and ensuring legal and regulatory compliance. With consumers' worries about food safety and provenance increasing, effective traceability systems provide a basis for building consumer confidence in food goods.

Supply chain management, or SCM, is the planning and coordinating of each stage involved in the production, distribution, and acquisition of food products. This comprehensive approach includes a variety of activities, including procurement, production scheduling, logistics, and inventory management. Streamlining supply chain procedures can help businesses enhance productivity, reduce costs, and improve the quality of their products. Furthermore, effective supply chain management (SCM) facilitates traceability efforts by ensuring that accurate and timely data about product movements and conditions is collected and shared throughout the supply chain.

In order to address the growing demand for sustainability and ethical sourcing, supply chain management and traceability must be integrated in addition to lowering risks associated with food safety and quality. Consumers are favoring products that can be tracked down and ethically supplied, which is pushing companies to use transparent and truthful practices everywhere. In order to meet market demands and legal obligations, this trend encourages companies to implement state-of-the-art

A. S. Jatoi (✉) · A. Iqbal · B. O. Nasir
Department of Chemical Engineering, Dawood University of Engineering and Technology, Karachi, Pakistan
e-mail: abdul.sattar@duet.edu.pk

A. S. Jatoi (ed.), *Integrated Food and Bioprocessing*,
https://doi.org/10.1007/978-3-031-87209-9_13

Fig. 13.1 Traceability in food supply chains. (Behnke & Janssen, 2020)

traceability systems and sustainable supply chain practices, which can increase their competitive edge.The intricate connection between supply chain management and traceability will be explored in this chapter, along with the strategies and tools that enable execution. We'll look at the legislative frameworks governing traceability, the challenges companies face in establishing dependable systems, and how customer perspectives affect these processes. Through case studies and best practices, we seek to illustrate the importance of supply chain management and traceability in creating an open, sustainable, and safe food system for all stakeholders (Fig. 13.1).

2 The Concept of Traceability

Traceability, as used in the food industry, is the ability to track and monitor food products' journey from raw materials to final consumption along the whole supply chain. Tracking from the point of origin to the end user is made possible by **forward traceability**, while tracking from the consumer back to the food product's source is made possible by backward traceability. The primary objective of traceability, which provides essential information on the origins, manufacturing, and distribution of food items, is to improve food safety and quality. To protect consumer trust and public health, this ability is essential for identifying and managing contamination, fraud, and product recall risks.

A key component of traceability is the gathering and administration of data across the supply chain. This report contains information on the procurement of ingredients, processing methods, storage conditions, and travel details. Because they make it simpler to precisely and effectively record the movements and transactions of items, technologies like barcodes, RFID, and blockchain are frequently employed in the installation of traceability systems. By maintaining accurate records, businesses can quickly identify the source of a food product and track it along the supply chain, enabling them to respond quickly to any potential safety or quality concerns.

Traceability is not only necessary for risk management, but it is also a helpful tool for complying with regulatory requirements and industry standards. Food businesses must put traceability systems in place in compliance with various regulations, such as those established by the Food and Drug Administration (FDA) in the US and the European Food Safety Authority (EFSA) in the EU, to ensure the safety and quality of food products. By adhering to these guidelines, businesses may boost consumer confidence in their brands and avoid costly penalties, product recalls, and damage to their reputation.

Along with regulatory compliance, traceability is increasingly recognized as a competitive differentiator in the market. Because they are more aware of the hazards to their food safety, customers today demand transparency on the management and sources of the products they buy. As a result, companies that prioritize traceability may see an increase in customer satisfaction and brand loyalty. By being transparent about their traceability practices, food firms can establish their brand, attract ethically conscious consumers, and take the lead in the growing market for safe, sustainable, and ethically produced food products.

3 Regulatory Frameworks and Standards

Establishing traceability standards in the food supply chain and ensuring the food products' safety, quality, and transparent origins depend on standards and regulatory frameworks. Laws requiring traceability processes have been implemented in several nations to protect public health and consumer interests. In the US, the Food and Drug Administration (FDA) has put in place initiatives such as the Food Safety Modernization Act (FSMA), which requires food makers to have traceability systems in place and emphasizes preventive controls. Similarly, the European Union's General Food Law requires all food companies to maintain effective traceability systems so that they can respond quickly to any food safety incident.

In addition to local regulations, international standards also control traceability practices in the food industry. Organizations such as the International Organization for Standardization (ISO) have developed standards such as ISO 22005 that provide guidelines for the development of traceability systems in food supply chains. These guidelines aim to enhance commercial relations and consumer safety by harmonizing practices across countries. Other industry-specific standards, like those set by the Global Food Safety Initiative (GFSI), also heavily rely on traceability, which encourages businesses to adopt best practices and maintain supply chain transparency.

The emphasis on traceability and regulatory framework compliance is further heightened by the growing consumer demand for transparency in food sourcing. As consumers become increasingly conscious of environmental sustainability and food safety, businesses must navigate a complex web of regulations and standards to meet both legal needs and customer expectations. By investing in robust traceability systems that adhere to regulatory standards, businesses may ensure compliance, boost consumer trust, foster brand loyalty, and lower risks associated with food safety and quality issues. Ultimately, adherence to these regulations is necessary to create a food supply chain that is safe, transparent, and resilient.

4 Technologies Supporting Traceability

The implementation of effective traceability systems in the food supply chain depends on modern technologies that enhance data collection, management, and communication. For traceability, one of the most used technologies is radio frequency identification, or RFID. RFID-tagged products can be traced automatically and in real time as they move through different stages of the supply chain. These tags store important product information, such as origin, processing details, and expiration dates, enabling companies to monitor inventory and ensure compliance with safety standards. RFID technology significantly reduces human error associated with manual data entry while improving product tracking speed and accuracy. Another crucial technology that enables traceability is **Blockchain**. From farm to fork, all supply chain transactions are recorded in a decentralized, immutable ledger system made possible by blockchain technology. Access to and verification of the data by all supply chain players guarantees transparency and accountability. This technology is highly helpful in increasing consumer trust since it allows consumers to follow the provenance of their food commodities, authenticate sustainability and sourcing claims, and ensure adherence to safety criteria. Furthermore, blockchain can speed up recall operations while reducing corporate expenses and public health hazards by enabling the timely identification of compromised batches (Fig. 13.2).

Additionally, cloud computing and data analytics play a major role in traceability activities. Because of the enormous amounts of data generated across the supply chain, businesses may utilize advanced analytics to learn about product movements, patterns, and potential risks. Stakeholder collaboration and information exchange are facilitated by cloud-based solutions, which make it simpler to combine data from several sources. By employing data analytics to inform decisions about inventory management, risk mitigation, and quality assurance, businesses may increase the effectiveness and efficiency of their traceability systems. The incorporation of these technologies will enhance food enterprises' capacity to guarantee supply chains' safety, quality, and transparency as technology develops further.

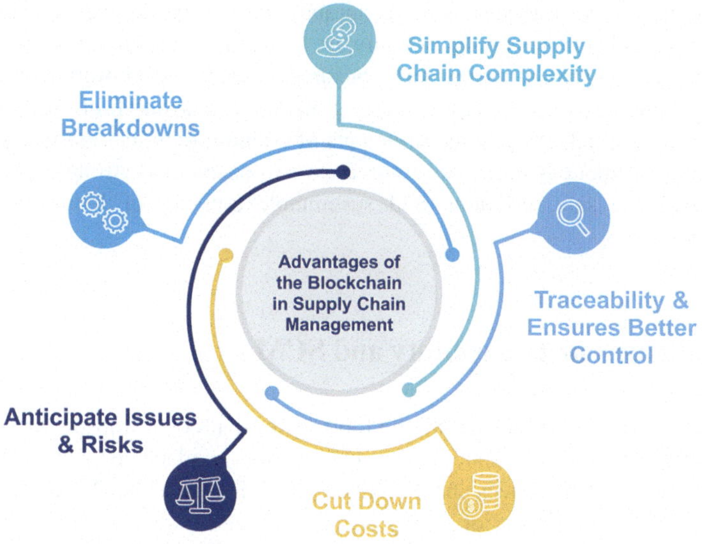

Fig. 13.2 Role of blockchain technology in supply chain management. (Narayanan et al., 2024)

5 Supply Chain Management Principles

Supply chain management, or SCM, encompasses a range of activities that facilitate the efficient flow of goods, information, and money throughout the food supply chain. SCM's main objective is to increase customer satisfaction through operational optimization while reducing risks and costs. One of the fundamental principles of supply chain management is collaboration among all parties involved, including producers, distributors, retailers, suppliers, and consumers. Businesses can share critical information, streamline processes, and better satisfy customer needs when they have strong partnerships and open lines of communication. Collaborating also enhances product traceability, maintaining supply chain transparency and providing relevant information to stakeholders.

The concept of integration is another essential SCM concept. This involves coordinating several activities, such as manufacturing, logistics, inventory management, and procurement, to ensure a seamless flow of goods and information. The ability of interconnected supply chains to respond faster to changing consumer preferences and market conditions leads to increased agility and resilience. For instance, the utilization of integrated technology platforms enables real-time inventory level monitoring, demand forecasting, and production scheduling, all of which help to minimize stockouts and excess inventory. This concept is especially important in the food industry, where timely delivery and product freshness are crucial to maintaining quality and customer satisfaction.

An other crucial SCM concept is continuous improvement. Organizations must regularly assess their performance and processes to identify areas for improvement and implement best practices. Two techniques that can be applied to cut waste, boost output, and enhance product quality are Lean and Six Sigma. By embracing a culture of continuous improvement, businesses can successfully innovate in response to client requests, adapt to shifting market trends, and fulfill legal requirements. In the end, adhering to the principles of collaboration, integration, and continuous improvement is necessary to create a responsive and strong supply chain that ensures the safe, efficient, and sustainable delivery of food products to consumers.

6 Challenges in Traceability and SCM

There are several challenges facing the food sector that could impede progress and compromise effectiveness and safety when putting in place robust supply chain management (SCM) practices and effective traceability systems. One of the biggest challenges is the complexity of supply chains. Among the several parties that regularly take part in modern food supply chains are farmers, processors, distributors, and retailers; each has their own protocols and processes. This intricacy can make it challenging to maintain correct and up-to-date records throughout the supply chain, which can also lead to issues with communication and data interchange. Inconsistent data formats and a lack of established protocols can further complicate tracing efforts, increasing the likelihood of errors and inefficiencies.

Another significant barrier is the cost of adopting and maintaining traceability systems. Strong traceability solutions require investments in technology, infrastructure, and training—all of which can be particularly costly for small and medium-sized enterprises. The costs of software systems, data administration, and RFID tags may deter businesses from fully committing to traceability projects. Additionally, ongoing expenses for system upkeep, regulatory compliance, and employee training can strain resources, especially for companies with thin profit margins. Because of this, many businesses struggle to balance the need to ensure product quality and safety with the financial implications of traceability.

Regulatory compliance is another problem that food companies face in relation to supply chain management and traceability. The food industry is subject to a number of laws and regulations that vary depending on the region and type of product. Following these regulations can be challenging, especially for companies with global operations. Noncompliance may result in severe fines, product recalls, and damage to one's reputation. Additionally, because regulations are always changing, businesses must keep up with the most recent developments and modify their procedures as necessary. The already difficult task of putting in place efficient supply chain management and traceability systems is made considerably more difficult by this changing atmosphere.

7 Case Studies and Best Practices

Successful applications of supply chain management (SCM) and traceability systems in the food industry provide valuable insights into workable strategies and solutions. One well-known example is Walmart, which has used blockchain technology to enhance its traceability efforts, particularly for fresh commodities. By collaborating with IBM to develop a blockchain-based system, Walmart will be able to track food supplies from farm to store. By using this technique, the company may quickly identify the source of foodborne illnesses and speed up recalls, ensuring the safety of its customers. Blockchain's transparency not only enhances traceability but also increases consumer trust in the Walmart brand, demonstrating how large retailers can leverage technology to improve food safety and supply chain efficiency.

Nestlé is another intriguing case study; it has implemented a comprehensive supply chain traceability system that prioritizes sustainable sourcing and quality control. Nestlé has developed a program called "Nestlé Cocoa Plan," to track cocoa beans from their origin to the final product. Nestlé supports the development of ethical farming practices and the fight against child labor by making sure that its chocolate is sourced responsibly. Through partnerships with farmers and mobile technology, Nestlé collects data on the agricultural and processing stages, facilitating continuous supply chain practice monitoring and enhancement. This commitment to traceability and ethical sourcing not only enhances product quality but also satisfies consumer needs for sustainability.

A third example is Tyson Foods, which has optimized its supply chain and enhanced traceability by utilizing cloud computing and data analytics. By integrating data from several sources, including as production, processing, and distribution, Tyson Foods can watch product movements in real-time, enhance inventory management, and reduce waste. The company also uses advanced analytics to predict demand and respond proactively to changing customer preferences. One example of best practices for maintaining efficiency while ensuring the quality and safety of its products is Tyson's focus on leveraging technology to enhance its supply chain capabilities. By employing innovative approaches to supply chain management and traceability, these companies show the potential benefits of investing in robust systems that enhance food safety, transparency, and consumer trust in the food sector.

8 Consumer Perspectives on Traceability

Consumer attitudes of traceability have significantly changed in recent years as a result of growing consumer awareness of food safety, quality, and ethical sourcing. Today's consumers are more informed and conscientious when choosing foods, often seeking out details regarding the products' origins and production processes. Because of this shift in consumer behavior, traceability is now a crucial factor to take into account when making purchases. Today's consumers demand brands to provide them with easily comprehensible information about the origin and processing of food products because they view traceability as a critical indicator of quality and safety.

Traceability is particularly being demanded by consumers who are worried about their health and want to be certain that the food they eat is safe and free of toxins. For these consumers, traceability systems that can quickly trace food products back to their source are crucial for identifying potential health risks and ensuring quality. By providing thorough information about the sources of materials and their production processes, businesses may boost consumer confidence and cultivate brand loyalty. Transparency in traceability reassures consumers that brands care about their health and safety, which encourages positive feedback and repeat business (Fig. 13.3).

Furthermore, ethical concerns have a big impact on what customers think about traceability. As consumers express worries about issues like environmental sustainability, fair labor standards, and animal welfare, they are increasingly searching for products that align with their values. Consumers can make educated purchases that are consistent with their beliefs by using traceability to verify claims regarding sustainable sourcing and ethical production. Consumers who respect morality are more inclined to identify with companies that embrace traceability and are transparent about their business practices, which will provide them a competitive advantage in the marketplace.

In conclusion, consumer perceptions of traceability highlight the need of transparency and accountability to the food supply chain. As consumers get more active in their food choices, they want businesses to provide accurate and honest information about the origins and handling of products. Setting traceability as a top priority not only helps companies win their customers' trust and loyalty, but it also positions them well in a market where safety, health, and ethical issues are becoming

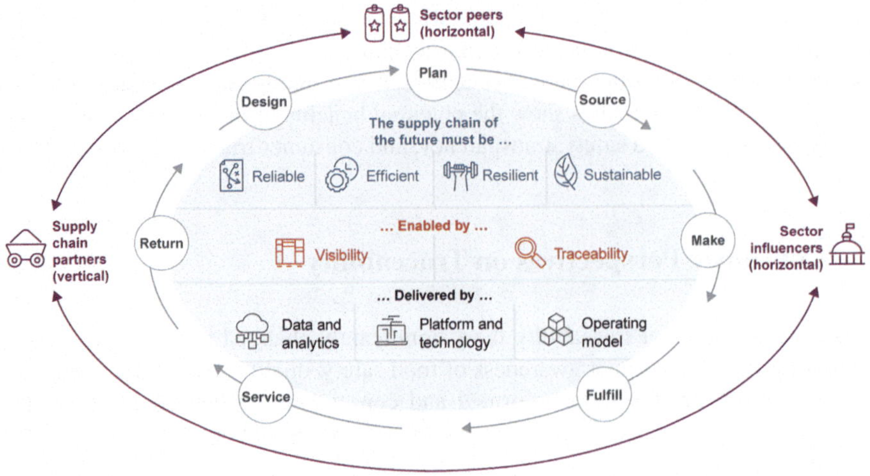

Fig. 13.3 Supply chain. (Narayanan et al., 2024)

increasingly significant. By effectively satisfying consumer demand for traceability, food firms can improve their connections with customers and contribute to the development of a safer, more sustainable food system.

9 Future Trends in Traceability and SCM

As the food industry continues to grow, several important trends in supply chain management (SCM) and traceability are emerging. These changes could alter how businesses operate and how consumers engage with food products. One of the most significant trends is the increasing adoption of **blockchain technology**. Accountability and transparency are enhanced by this decentralized ledger technology, which provides an unchangeable record of every transaction across the supply chain. To track products from farm to table, businesses are increasingly investigating blockchain technologies. This enables them to quickly identify the sources of contamination, verify sustainability claims, and boost overall customer confidence. As blockchain technology advances, it is expected to become standard practice for ensuring traceability and building strong supply chains.

Another trend is the integration of Internet of Things (IoT) devices and smart sensors into supply chain processes. IoT technology enables real-time monitoring of environmental parameters including temperature, humidity, and location throughout the transportation and storage of food products. By collecting and analyzing this data, businesses can ensure that products are stored in optimal conditions, reducing spoiling and enhancing quality. The widespread usage of smart devices allows for more precise product tracking and tracing and allows for the timely resolution of potential issues. IoT technology's use in traceability and supply chain management is expected to grow as it becomes more widely available and reasonably priced, lAs IoT technology becomes more accessible and affordable, its application in supply chain management and traceability is anticipated to increase, resulting in further improvements in food safety and efficiency. Reading to additional efficiencies and advancements in food safety.

Sustainability will remain a top focus for supply chain management and traceability. Businesses are being forced to adopt more sustainable supply chain practices as a result of the growing demand for products that are ethically and environmentally sourced. This trend is pushing companies to implement comprehensive traceability systems that monitor the quality of their products as well as the environmental impact of their activities. Traceability programs that integrate sustainable sourcing and production methods will meet legal requirements, CSR goals, and customer desires. As sustainability becomes a crucial component of corporate strategy, traceability will be crucial in demonstrating a company's commitment to moral business conduct and fostering client loyalty.

In conclusion, shifting consumer demands and technology advancements will likely bring about significant changes to supply chain management and traceability in the future. Blockchain, IoT, and sustainable practices will totally transform how businesses handle accountability, transparency, and food safety. By staying ahead of these trends, businesses may increase their operational efficiency, win over customers, and contribute to the development of a more resilient and sustainable food system.

10 Conclusion

In conclusion, traceability and supply chain management are essential components of the modern food industry that are necessary to ensure food safety, quality, and transparency. Companies must implement dependable traceability systems and streamline their supply chain processes in response to consumers' increasing needs for assurance regarding the origin and handling of their food. Integrating cutting-edge technologies like blockchain, IoT, and data analytics could increase the effectiveness and efficiency of current systems by enabling real-time supply chain tracking, monitoring, and accountability.

The evolving regulatory landscape emphasizes even more how important compliance and best practices are to supply chain management and traceability. As regulations tighten, businesses that prioritize transparency and adhere to industry standards will not only lower risks but also gain a competitive advantage in the marketplace. Addressing ethical and sustainable concerns in sourcing and manufacturing practices will also appeal to consumers who are increasingly searching for companies that share their values.

The future of traceability and supply chain management will ultimately depend on how well players along the food supply chain embrace innovation and collaboration. By forming alliances, investing in technology, and embracing sustainable practices, businesses can increase their resilience and adaptability in a rapidly changing environment. As the food industry grows, effective traceability and supply chain management will be essential to gaining consumer trust, ensuring food safety, and creating a more sustainable and accountable food system for all.

References

Behnke, K., & Janssen, M. (2020). Boundary conditions for traceability in food supply chains using blockchain technology. *International Journal of Information Management, 52*, 101969.

Narayanan, G., I. Cvitić, D. Peraković, & S. Raja (2024). Role of blockchain technology in supplychain management. *IEEE Access*.

Chapter 14
Challenges, Ethical Considerations, and Future Trends

Abdul Sattar Jatoi, Arshad Iqbal, Bazil Owais Nasir, and Owais Baig

1 Introduction

Numerous problems and ethical dilemmas have emerged in a variety of disciplines as a result of the quickening pace of technological development and the growing complexity of societal issues. Organizations must consider the effects of their choices as they work to innovate and adapt to the needs of a changing environment. This chapter attempts to break down these difficulties, offering a thorough summary of the difficulties practitioners encounter in the fast-paced world of today. The many variables affecting decision-making, from negotiating regulatory frameworks to guaranteeing public acceptance, are essential to comprehending the field's future course.

Public policies and organizational practices are significantly shaped by ethical issues. The ethical conundrums pertaining to social justice, environmental impact, and data privacy change along with technology. Stakeholders are being asked to strike a balance between their moral obligations to society and the environment and the pursuit of innovation. This chapter promotes a comprehensive strategy that takes into account both profit and purpose, highlighting the significance of ethical frameworks and discussions that direct decision-making processes.

Looking ahead, new trends and best practices will be shaped by the interaction of ethical issues and obstacles. Stakeholders can better predict changes in consumer preferences, legislative changes, and technical improvements by having a deeper awareness of the existing landscape. This chapter provides a starting point for investigating these patterns, giving readers the knowledge they need to handle the changing intricacies of their specialized industries. By this investigation, we hope to

A. S. Jatoi (✉) · A. Iqbal · B. O. Nasir · O. Baig
Department of Chemical Engineering, Dawood University of Engineering and Technology, Karachi, Pakistan
e-mail: abdul.sattar@duet.edu.pk

A. S. Jatoi (ed.), *Integrated Food and Bioprocessing*,
https://doi.org/10.1007/978-3-031-87209-9_14

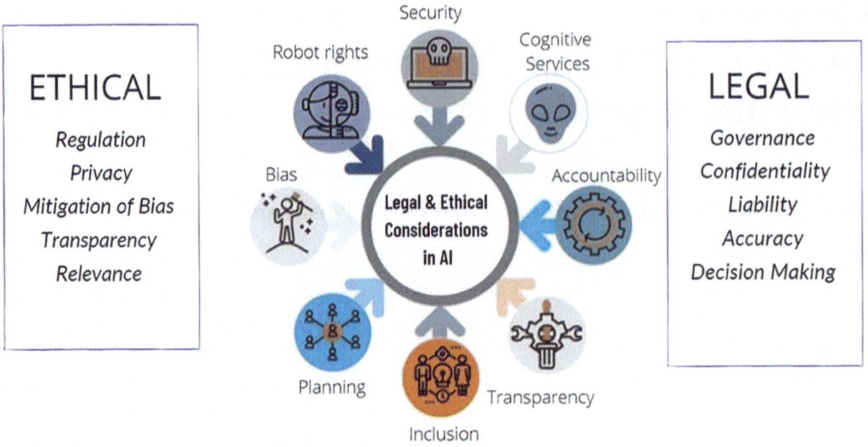

Fig 14.1 Legal and ethical consideration in artificial intelligence in healthcare. (Naik et al., 2022)

promote a more accountable and inclusive approach to development, guaranteeing advancement without sacrificing moral rectitude or social duty. This chapter explores the ethical ramifications of the field's decision-making procedures in addition to addressing difficulties. It takes into account ethical obligations, data privacy issues, and environmental ethics, promoting a well-rounded strategy that puts sustainabilitySustainability and stakeholder well-being first. The chapter looks ahead and analyzes trends that will influence the landscape, such as changes in market dynamics and technology breakthroughs. It seeks to give stakeholders the resources they need to successfully and morally traverse this changing environment by putting forward best practices and cooperative tactics (Fig. 14.1).

2 Current Challenges

Many industries encounter numerous obstacles that hamper advancement and creativity in the fast-paced, globally interconnected world of today. For stakeholders trying to negotiate the complexities of their respective areas, it is essential to comprehend these challenges. The difficulties of today can be roughly divided into four categories: social, economic, regulatory, and technology.

1. Technological Challenges

Organizations face integration challenges as technology continues to develop at an unprecedented rate. It is frequently necessary to align new systems with pre-existing infrastructures, which might result in operational inefficiencies and higher expenses. Additionally, scalability is a major challenge; the influence of many creative ideas is limited since they are difficult to adapt to wider contexts. Additionally, it might be resource-intensive to spend in training and constant learning to stay up with the rapid developments in technology.

2. Regulatory Challenges

Another major obstacle is navigating the complicated regulatory environment. It can be difficult to comply with local, national, and international rules, especially for businesses that have operations in several different countries. Regulatory frameworks frequently lag behind technical developments, creating loopholes that may impede innovation or put businesses at danger from the law. Additionally, companies are put under more stress because it takes specialized resources to stay up to date with changing legislation.

3. Economic Challenges

Operational viability is also significantly impacted by the state of the economy. The ability of an organization to maintain growth and innovate can be greatly impacted by changes in funding, resource allocation, and market rivalry. Since businesses must strike a balance between investing in new technology and current economic pressures, financial sustainability is a top concern in many industries. Furthermore, consumer demand may be impacted by market volatility, creating uncertainty that makes long-term planning more difficult.

4. Social Challenges

Public opinion and social approval are essential elements of a successful deployment. Public skepticism or opposition to many technologies stems from worries about privacy, safety, and ethical ramifications. Furthermore, because marginalized communities frequently do not have fair access to new technology and solutions, guaranteeing accessibility and inclusivity continues to be a critical concern. To achieve broad acceptability, organizations must manage these social factors in order to cultivate engagement and trust with their stakeholders.

Stakeholders can improve their ability to negotiate the complexity of their environments by acknowledging and resolving these issues. In the parts that follow, we will examine the moral issues raised by these difficulties and look at ways to proceed in a responsible and efficient way.

3 Ethical Considerations

As businesses face a variety of difficulties in the complicated world of today, ethical issues have become important determinants of operational procedures and decision-making. Ethical frameworks offer direction for handling conundrums brought on by social obligations, legal compliance, and technology breakthroughs. The main ethical issues that stakeholders need to solve in order to guarantee ethical and sustainable behaviors are listed in this section (Fig. 14.2).

Fig 14.2 Ethical consideration (Naik et al., 2022)

1. Moral Implications

Organizations have moral obligations that go beyond following the law; they also include considering how decisions may affect stakeholders, communities, and the environment more broadly. Businesses are being asked more and more to think about the long-term effects of their decisions, aiming for results that advance well-being rather than cause harm. This involves evaluating how new technology will affect society, including whether they will increase inequality or cause job displacement. Fostering responsible decision-making in businesses requires establishing a culture of ethical awareness and accountability.

2. Data Privacy and Security

Concerns about data security and privacy have become critical in a time when data is a valuable commodity. The ethical ramifications of gathering, preserving, and using personal data must be managed by organizations. It can be difficult to strike a balance between the need for data-driven insights and respect for people's right to privacy. To safeguard sensitive data and maintain transparency on data usage methods, stakeholders must put strong cybersecurity safeguards in place. Furthermore, the use of AI and machine learning raises ethical questions since biased algorithms might produce unjust results that call for constant review and modification.

3. Environmental Ethics

It is impossible to ignore the moral implications of environmental sustainability. It is the duty of organizations to reduce their ecological footprint and take into account how their actions may affect natural resources in the long run. This entails dealing with problems including waste management, biodiversity preservation, and carbon emissions. Organizations should use ethical principles when adopting sustainable practices that improve environmental health in addition to adhering to legal requirements. Putting a focus on corporate social responsibility (CSR) programs helps improve moral interaction with environmental concerns.

4. Equity and Justice

A basic ethical responsibility for organizations functioning in varied and complex situations is ensuring equity and justice. This entails identifying and resolving inequalities among stakeholders and within communities. Organizations must actively seek inclusive practices that empower oppressed groups and advance social justice in order to make ethical decisions. This can include programs designed to make goods and services more accessible as well as to promote inclusive and diverse work environments. For their procedures to have an equal impact on all demographics, organizations must be proactive in detecting and reducing any biases that may be present.

Organizations can improve their reputations, gain the trust of stakeholders, and benefit society by incorporating ethical issues into their basic strategies. In order to change their practices to reflect changing society norms and expectations, stakeholders must have continuous conversations on ethics as we look to the future. We shall examine upcoming trends that will influence the environment and their consequences for moral judgment in the years to come in the next part.

4 Future Trends

As the twenty-first century progresses, a number of trends have the potential to influence how businesses and organizations function, develop, and interact with stakeholders. To be ready for the opportunities and challenges that lie ahead, it is imperative to comprehend these patterns. This section outlines a number of significant emerging developments that are anticipated to have an influence on operational procedures and ethical issues.

1. Technological Advancements

Change is still being driven by rapid technology innovation in all industries. New technologies like blockchain, artificial intelligence, machine learning, and the Internet of Things (IoT) are changing business models and operational capacities. These developments bring with them both possibilities and moral conundrums. While using these technologies to boost productivity and creativity, organizations must manage the effects they may have on privacy, security, and job displacement. The emphasis will gradually move to responsible tech adoption, making sure that innovations are applied in a morally and openly responsible manner.

2. Regulatory Changes

Regulatory frameworks will change along with technology. Legislative changes may aim to address new ethical issues pertaining to cybersecurity, data privacy, and environmental sustainability. Businesses need to keep up with these changes in regulations and modify their procedures as necessary. Developing regulations that strike a balance between innovation, public safety, and ethical norms will need active collaboration with stakeholders and lawmakers. Establishing international standards that direct ethical behavior across borders will require cooperation.

3. Market Trends

Growing knowledge of social and environmental issues is causing a rapid shift in consumer choices. Organizations that value sustainability, transparency, and ethical practices are likely to benefit from future market trends. Corporate social responsibility (CSR) is a crucial part of business strategy since consumers are becoming more discriminating and will support brands that share their beliefs. Businesses that successfully incorporate sustainability into their operations and exhibit a dedication to moral conduct will have a stronger chance of prospering in cutthroat marketplaces.

4. Social Trends

Organizational procedures and cultural norms will continue to be influenced by the increased focus on equity and social justice. As diversity, equity, and inclusion (DEI) movements gain traction, corporations are being forced to review their procedures and policies. A greater emphasis on developing inclusive workplaces, encouraging diverse leadership, and making sure that goods and services satisfy the requirements of all communities are probably going to be some of the future trends. Companies must interact with stakeholders to learn about their issues and proactively seek solutions that advance equality and justice.

5. Focus on Sustainability

Due to pressure from regulations and consumer demand, sustainability is becoming a more important factor for businesses. Future developments will highlight how important it is for businesses to implement sustainable practices that reduce their negative effects on the environment and advance social welfare. This entails switching to renewable energy, embracing the ideas of the circular economy, and putting responsible sourcing into effect. Businesses will need to link their goals with social responsibility and environmental stewardship as sustainability becomes a fundamental part of their company identity (Fig. 14.3).

Organizations can successfully traverse the changing terrain while maintaining ethical standards by identifying and adjusting to these future developments. Long-term success and societal influence will depend on embracing innovation, encouraging inclusion, and placing a high priority on sustainability. We will stress the significance of continued engagement with difficulties, ethical considerations, and new trends in the concluding part, which will also provide a summary of the main insights presented.

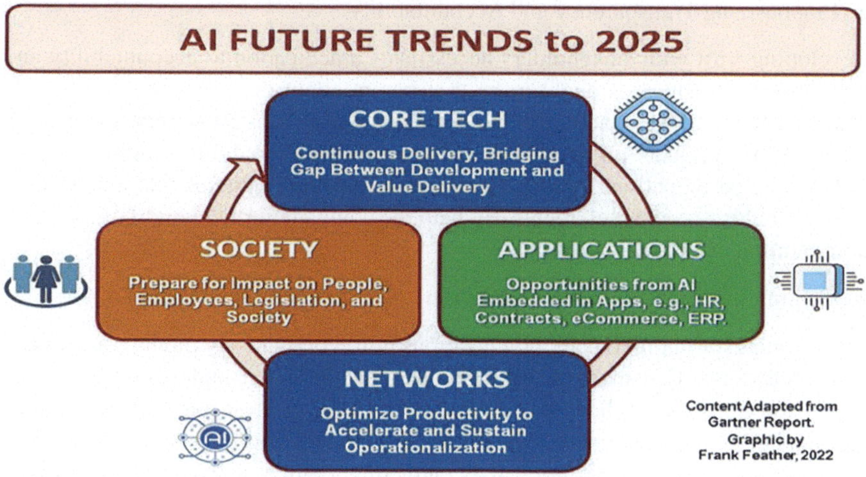

Fig 14.3 Trends in AI and its impact on society. (Pavaloiu, 2016)

5 Strategies for Addressing Challenges and Ethical Issues

Adopting successful solutions becomes essential for promoting ethical practices and sustainable growth as organizations negotiate the complexity of contemporary issues and ethical considerations. Several important tactics that can assist companies in overcoming these obstacles and maintaining moral principles are described in this section.

1. Best Practices for Ethical Decision-Making

Establishing a strong ethical framework inside businesses is crucial for directing decision-making. This may entail creating a code of ethics that delineates fundamental ideals and beliefs, serving as a point of reference for staff members. A culture of ethical awareness and accountability can be promoted by frequent training sessions and workshops on ethical practices. Employees can be empowered to make responsible decisions by fostering candid conversations about moral conundrums and providing secure avenues for raising issues.

2. Collaboration and Partnerships

Working together with stakeholders, such as other businesses, government agencies, and neighborhood associations, can improve how well problems and moral dilemmas are handled. Through multi-stakeholder engagement, organizations can work together to address shared issues and exchange best practices, resources, and information. Partnerships with advocacy organizations and NGOs, for instance, might offer insightful information on ethical issues and community needs. Working together can also make it easier to follow rules and increase operational openness.

3. Emphasizing Transparency and Accountability

Developing trust with stakeholders necessitates a dedication to accountability and transparency. Businesses ought to be transparent about their procedures, choices, and the effects of their activities. Reporting on social responsibility, ethics, and sustainability projects on a regular basis can increase credibility and encourage stakeholder participation. Putting in place accountability systems like independent audits and third-party assessments can help the company further solidify its dedication to moral behavior.

4. Continuous Monitoring and Evaluation

To find areas for improvement, organizations must continuously monitor and assess their procedures. Organizational performance can be better understood by implementing performance indicators pertaining to social impact, sustainability, and ethical standards. Organizations can maintain their agility and responsiveness to new problems and moral quandaries by conducting regular assessments. Involving stakeholders in feedback loops can help guarantee that practices conform to changing societal norms and community expectations.

5. Education and Awareness Campaigns

It is essential to educate stakeholders and employees about ethical issues and the value of responsible behavior. Workshops, seminars, and awareness campaigns that draw attention to ethical concerns and their ramifications are examples of educational programs. Giving staff members the means and instruments to make moral decisions can enable them to make wise decisions. Outreach initiatives that involve the community can also help people understand ethical issues more broadly and work together to find solutions.

6. Leveraging Technology for Ethical Compliance

Technological developments can be very helpful in resolving ethical problems and dilemmas. Organizations can more efficiently monitor adherence to laws and ethical standards by putting data analytics solutions into place. Blockchain and other technologies can increase supply chain transparency, guaranteeing ethical and responsible sourcing. Organizations can strengthen their capacity to overcome obstacles while maintaining moral integrity by utilizing creative solutions.

By implementing these tactics, businesses can proactively handle difficulties and moral dilemmas while promoting an accountable and responsible culture. A dedication to moral behavior and stakeholder involvement will be essential as they negotiate the complexity of their surroundings in order to achieve long-term success and have a beneficial social influence. The chapter's findings will be summed up in the final part, which will also highlight the significance of continuing to be ethically conscious in future undertaking

6 Global Perspectives

Global perspectives on integrated food and bioprocessing highlight the growing importance of biotechnology in the development of sustainable food production systems. With increasing global populations and the challenges posed by climate change, traditional agricultural and food production methods face limitations in meeting the growing demand for nutritious food. Integrated food and bioprocessing approaches combine biological, chemical, and engineering principles to improve the efficiency and sustainability of food production. These methods emphasize the use of biotechnological processes such as fermentation, enzyme applications, and microbial technologies to enhance food quality, shelf life, and nutritional value while reducing waste and energy consumption. Countries worldwide are adopting these methods to address food security and sustainability issues, promoting innovation in food processing industries. On a global scale, the integration of food and bioprocessing also reflects the need for cross-border collaboration and the sharing of knowledge to solve complex challenges. Different regions are leveraging local resources and technologies, adapting bioprocessing techniques to suit their unique agricultural conditions and dietary needs. For instance, in developing countries, biotechnologies are used to enhance staple food crops, improve food safety, and reduce post-harvest losses, which are critical to combating hunger and malnutrition. Simultaneously, developed nations are advancing in the creation of plant-based proteins and other alternative food sources, further pushing the boundaries of integrated bioprocessing. As food systems continue to evolve, global cooperation, regulatory alignment, and the adoption of ethical and sustainable practices will be essential for maximizing the potential of integrated food and bioprocessing technologies.

7 Economical Impact

The economic impact of integrated food and bioprocessing technologies is profound, influencing various sectors including agriculture, manufacturing, biotechnology, and retail. By improving the efficiency of food production and reducing waste, these technologies contribute significantly to economic growth, especially in developing countries. Bioprocessing methods such as fermentation, enzyme applications, and waste-to-resource technologies help lower production costs, create value-added products, and enhance the sustainability of food production. This, in turn, allows businesses to meet consumer demands more efficiently, fostering growth in both domestic and international markets. Additionally, the increasing focus on sustainable practices through bioprocessing can result in long-term economic benefits by preserving resources and mitigating environmental damage, which could otherwise lead to higher costs associated with resource depletion or environmental cleanup. Furthermore, the bioprocessing industry opens up opportunities for job creation and innovation in several industries. Biotechnology and food processing sectors are thriving due to the demand for skilled professionals in research, development, and production. This boost in employment extends to the creation of new markets, such as plant-based foods, biofuels, and bio-based chemicals, which are gaining traction due to their environmentally friendly attributes. By investing in bioprocessing, governments and businesses can stimulate local economies and create new revenue streams while supporting agricultural sustainability and improving food security. The continued growth of integrated food and bioprocessing technologies is likely to remain a key driver of economic advancement globally, fostering a resilient, sustainable food system that can withstand future economic and environmental challenges.

8 Conclusion

The difficulties and moral dilemmas that companies face are greater than ever in a world that is becoming more linked and complex. The complex web of technological, legal, financial, and social issues that influence decision-making has been examined in this chapter. Organizations must confront the moral ramifications of their acts and interact with the larger ethical frameworks that govern responsible practices as they attempt to innovate and satisfy the needs of changing markets.

The ethical considerations discussed—ranging from data privacy and environmental responsibility to equity and social justice—underscore the importance of integrating ethical principles into the core operations of organizations.

Without a doubt, new dynamics and trends will emerge in the future, reshaping the landscape of many different professions. Organizations can put themselves in a position to prosper in this changing climate by proactively embracing innovation, encouraging cooperation, and making a commitment to sustainability. In the end, incorporating ethical considerations into business operations is not only morally required, but also advantageous from a strategic standpoint, with the potential to result in long-term growth and beneficial social effects. A shared dedication to moral participation will be essential in shaping a responsible and just future as stakeholders continue to negotiate these issues.

References

Naik, N., Hameed, B. Z., Shetty, D. K., Swain, D., Shah, M., Paul, R., Aggarwal, K., Ibrahim, S., Patil, V., & Smriti, K. (2022). Legal and ethical consideration in artificial intelligence in healthcare: who takes responsibility? *Frontiers in Surgery, 9*, 862322.

Pavaloiu, A. (2016). The impact of artificial intelligence on global trends. *Journal of Multidisciplinary Developments, 1*(1), 21–37.

Chapter 15
The Future of Integrated Food and Bioprocessing

Abdul Sattar Jatoi, Arshad Iqbal, Bazil Owais Nasir, and Owais Baig

1 Introduction

Over the past few decades, the integrated food and bioprocessing business has witnessed enormous transformation due to technology developments, changes in client tastes, and rising awareness of environmental issues. The objective of this chapter is to evaluate the field's future while emphasizing how food processing and bioprocessing are related. As sectors grapple with the challenges of population growth and climate change, there is an increasing need for innovative, efficient, and sustainable approaches. This introductory section sets the stage for understanding how integrated approaches might address these global concerns.

Sustainability is a recurrent theme in this chapter. In addition to the environmental impacts of food production, it encompasses the social responsibility and economic sustainability of food systems. By examining the relationships between sustainability objectives and technological advancements, the chapter aims to provide a comprehensive knowledge of how integrated food and bioprocessing might contribute to a more resilient food system. The significance of adapting to these changes cannot be overstated, as the future of food production hinges on our ability to innovate while preserving environmental sensitivity.

As we move through the chapter, a range of trends, technologies, and challenges will be looked at to give an idea of what the future might hold. Integration of food and bioprocessing systems will be one of the primary topics, emphasizing how these fields may work together to boost output, reduce waste, and improve the overall quality and safety of food (Fig. 15.1).

A. S. Jatoi (✉) · A. Iqbal · B. O. Nasir · O. Baig
Department of Chemical Engineering, Dawood University of Engineering and Technology, Karachi, Pakistan
e-mail: abdul.sattar@duet.edu.pk

A. S. Jatoi (ed.), *Integrated Food and Bioprocessing*,
https://doi.org/10.1007/978-3-031-87209-9_15

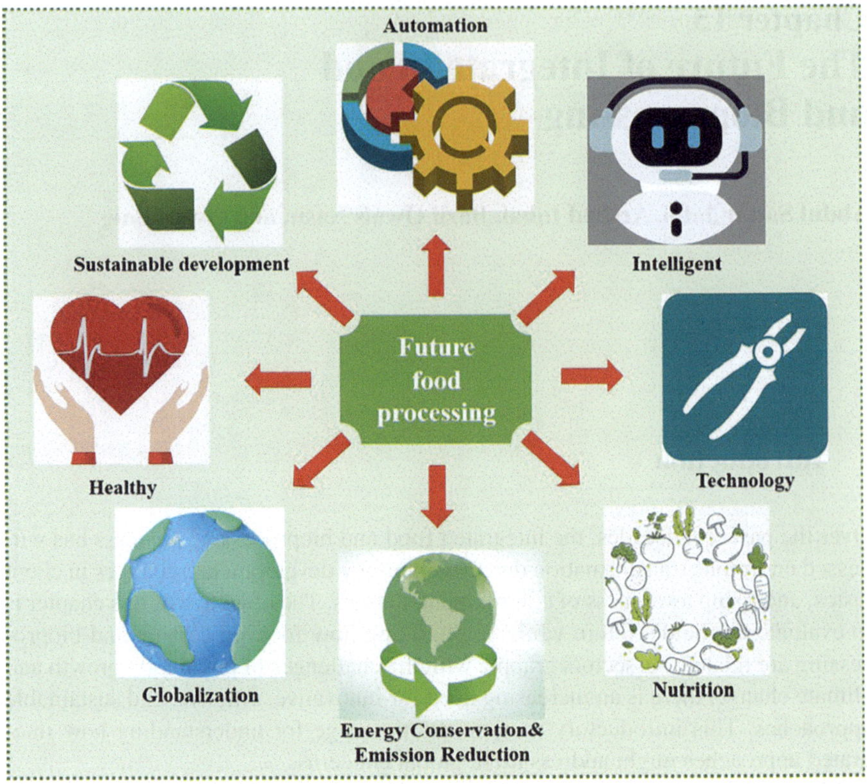

Fig. 15.1 Novel drying technologies for future animal- and plant-derived protein foods: research progress, challenges, and potential applications. (Niu et al., 2024)

2 Technological Advancements

Technological developments are at the vanguard of the revolution in integrated food and bioprocessing. Automation and robots are two examples of new technology that are increasing accuracy, reducing labor costs, and streamlining industrial operations. Automated systems that monitor and modify production parameters in real time can reduce waste and ensure consistent product quality. Additionally, production scheduling and inventory management are optimized by predictive analytics enabled by advancements in artificial intelligence and machine learning, which boost output and reduce operating costs.

Furthermore, biotechnology is becoming more and more significant in food and bioprocessing. Genetic engineering and synthetic biology can be used to produce crops with enhanced traits, such as increased resistance to pests or higher nutritional profiles. These developments can help answer concerns about food security while lessening the adverse environmental effects of traditional farming practices. The innovative applications of fermentation technology outside traditional food

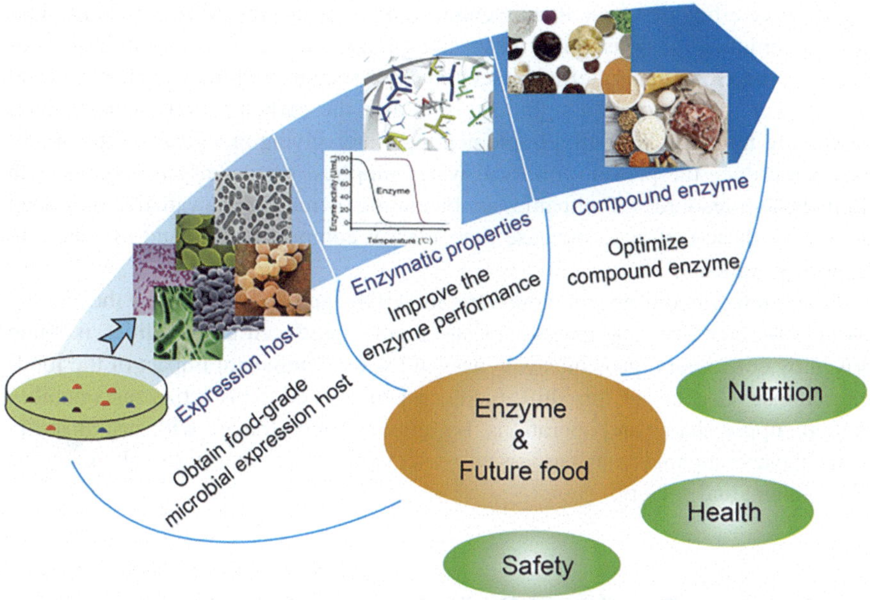

Fig. 15.2 Current progress and prospects of enzyme technologies in future foods. (Pang et al., 2021)

products, like the production of biofuels, bioplastics, and other sustainable materials, further highlight the versatility of bioprocessing methods.

The integration of the Internet of Things (IoT) into the food processing industry is transforming the collection and analysis of data. Smart sensors can track a number of attributes during the production process, providing useful data that enhances decision-making. These technologies work together to create a more cohesive system. Facilitating the seamless flow of data between distribution, processing, and production. This enhances the food supply chain's overall sustainability and effectiveness (Fig. 15.2).

3 Sustainability Initiatives

Sustainability practices are becoming more and more important in integrated food and bioprocessing. One crucial tactic is the notion of a circular economy, which reduces waste and values byproducts. Waste management strategies need to be reconsidered to ensure that resources are recycled, reused, or repurposed. For instance, food waste can be converted into energy by anaerobic digestion. Additionally, it can be used as a feedstock for animal feed or biofertilizers. This reduces the environmental effect and increases resource efficiency.

Resource efficiency is another essential element of sustainability in food production. Techniques for conserving water and energy are growing in popularity. For example, the use of energy-efficient technologies such as heat recovery systems and solar power integration can significantly reduce the carbon footprint of the food processing industry. Similarly, implementing water-saving practices and technologies can reduce the pressure on local water supplies, particularly in regions with limited water resources. Companies are becoming increasingly aware that they need to employ these tactics to increase their market competitiveness and to adhere to legal requirements.

Future integrated food and bioprocessing will be greatly impacted by the emphasis on sustainability. The market for sustainably produced commodities is being driven by consumers' growing awareness of the environmental impact of the foods they choose. Because of this shift, food companies need to prioritize sustainability in their supply chains and operations. By aligning their strategy with sustainability goals, these companies may surpass customer expectations and positively affect the environment and society.

4 Consumer Trends and Preferences

Understanding consumer trends is critical to the future success of integrated food and bioprocessing. One such trend is the increase in demand for products with a nutrition and health focus. Consumers who are more concerned about their health are searching for functional meals that offer specific health benefits, such improved immune system or digestive support. In addition to providing unambiguous information on the nutritional content and origins of their goods, this movement is pushing food producers to be creative and include health-promoting substances.

Ethical consumption is another growing consumer preference that is impacting future developments in food and bioprocessing. Growing interest in plant-based diets and alternative proteins is a result of worries about animal abuse and the environmental impact of livestock farming. As a result, companies are developing innovative plant-based goods and meat substitutes that appeal to these consumers. This shift not only creates new markets but also forces established food producers to adapt their goods to meet the evolving preferences of their customers.

Product development and marketing strategies need to take consumer preferences into account if companies want to thrive in the competitive food sector. As consumers seek out items that align with their beliefs, companies that prioritize sustainability, transparency, and health are likely to see greater success. Businesses will need to continuously innovate and adapt to changing consumer demands, and the ability of integrated food and bioprocessing to handle these changes will decide its future.

5 Regulatory and Policy Framework

The regulatory landscape surrounding integrated food and bioprocessing is evolving along with technology and consumer expectations. Governments everywhere are emphasizing the importance of ensuring food safety, quality, and sustainability through comprehensive legislation. This section will outline the current regulatory frameworks that control food processing and bioprocessing and analyze how they impact innovation and production techniques.

Future regulations are likely to highlight the importance of traceability and transparency in food supply chains. Customers are calling for more information on where their food comes from and how it is produced. Regulatory bodies may respond by passing stricter food safety laws and labeling specifications in an effort to increase customer confidence. As science advances, regulations relating to genetically modified organisms (GMOs) and new food products will also need to be carefully examined in order to balance safety concerns with the need for innovation.

Collaboration between governments, industry stakeholders, and research institutions will have a significant impact on the regulatory environment of the future. Lawmakers must work closely with food growers and processors to create frameworks that support innovation while preserving public health and safety. By including stakeholders in the regulation process, governments may create an atmosphere that supports technical advancements and sustainable practices in the food and bioprocessing sectors.

5.1 Shifting Proteins: From Conventional to Sustainable Alternatives for a Healthier Future

Shifting Proteins: From Conventional to Sustainable Alternatives for a Healthier Future explores the evolving landscape of protein sources as the world grapples with pressing challenges like environmental degradation, health concerns, and sustainability in food production. Traditional protein sources, such as meat and fish, have long been staples in global diets but come with significant drawbacks. These include high environmental costs, such as greenhouse gas emissions, land and water usage, and overfishing, as well as health issues linked to excessive consumption of animal-based proteins. As the global population grows and demands for protein increase, these conventional sources are becoming less viable, pushing the need for alternatives that are more sustainable and less harmful to the planet. The shift toward non-conventional protein sources, such as plant-based proteins, insects, and algae, offers promising solutions. These alternatives provide high nutritional value while drastically reducing environmental impacts, including lower carbon footprints, water usage, and land requirements. As innovation in food technology continues, non-conventional proteins are becoming increasingly accessible and acceptable to consumers as higlighted in figure 15.3. Not only do they help meet the rising global

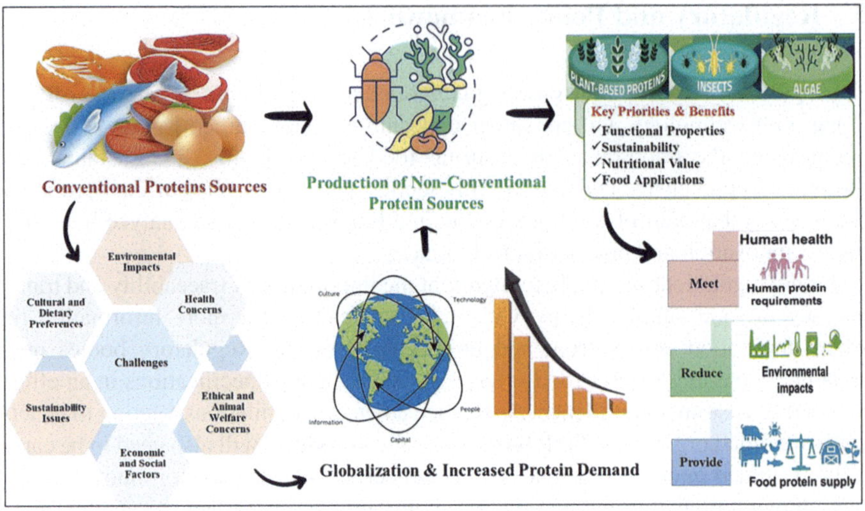

Fig 15.3 Shifting Proteins (Zhang W, Liu C, Zhao J, Ma T, He Z, Huang M, Wang Y. Modification of structure and functionalities of ginkgo seed proteins by pH-shifting treatment. Food chemistry. 2021 Oct 1;358:129862.)

demand for protein, but they also offer a healthier, more sustainable way forward, addressing both nutritional needs and environmental concerns. With the growing acceptance of these alternatives, supported by research and consumer education, we are moving toward a future where food systems can sustain both human health and the planet's resources.

6 Market Dynamics and Economic Factors

Global demand for integrated food and bioprocessing is influenced by a number of economic factors, including supply chain dynamics, international trade regulations, and consumer demand. Customers' increased awareness of sustainability and health issues has led to a substantial growth in the business in recent years. This section will look at current market trends, highlighting the industry's financial opportunities and challenges.

Companies must make an investment in innovation if they want to capitalize on emerging market trends. Funding for research and development of food and bioprocessing technologies is essential to promoting innovation and increasing competitiveness. This section will discuss how government grants, venture capital, and public-private partnerships support industry innovation. Companies that prioritize research and development can produce innovative products and processes that meet customer and legal requirements.

Globalization is also causing changes in the sector of integrated food and bioprocessing. Companies must manage global trade while dealing with opportunities and challenges as markets become increasingly interconnected. Access to a global supply chain can expand market reach and resource availability, but it also entails risks of supply chain disruptions and fluctuations in commodity prices. Businesses looking to thrive in the cutthroat global economy must comprehend these dynamics.

7 Challenges and Risks

Despite the many advantages of integrated food and bioprocessing, there are a number of risks and challenges to take into account. Integrating new technology into existing production processes is one of the technological challenges. Many food processors find it challenging to adopt modern automation, IoT, and data analytics due to implementation complexity, costs, and worker training needs. As companies collect and analyze enormous amounts of operational and consumer data, worries about data security and privacy are also becoming more pressing.

Additionally, market uncertainties pose significant challenges for the sector. Changes in consumer demand may lead to variations in production levels and profitability. To minimize potential losses, companies need to be adaptable and aware of changing market conditions. Supply chain disruptions, which are exacerbated by global events such as pandemics or geopolitical crises, further complicate the operational environment. Businesses must develop robust risk management plans in order to effectively manage these risks.

To overcome these challenges, collaboration and innovation across the industry will be required. By forming alliances, businesses can take advantage of the collective knowledge of stakeholders, including technology suppliers, academic institutions, and government regulators, to mitigate these risks. The future of the integrated food and bioprocessing industry will depend on its capacity to overcome challenges and take advantage of opportunities for sustained growth.

8 Case Studies and Best Practices

Real-world examples of successful integrated food and bioprocessing may provide valuable insights into creative and successful strategies. This section will provide case studies of companies that have successfully applied integrated strategies to increase sustainability, productivity, and product quality. These illustrations will demonstrate how companies have seized opportunities and surmounted challenges in the ever-evolving landscape.

We'll highlight the importance of collaboration, innovation, and adaptability while showcasing best practices in integrated food and bioprocessing. We'll discuss companies that have embraced the principles of the circular economy, including those that employ waste reduction and resource recovery strategies.

9 Challenges in Scaling and Implementation

Scaling and implementing integrated food and bioprocessing technologies present a variety of challenges, primarily related to technological, economic, and environmental factors. One of the main obstacles is the technological limitations associated with scaling up bioprocessing techniques from laboratory or pilot-scale operations to large-scale industrial production. In many cases, biotechnological processes that work effectively in small batches may not perform efficiently when applied on a larger scale, often due to the complexity of maintaining consistent conditions, optimizing microbial growth, and controlling reaction variables. Ensuring the reproducibility and reliability of these processes at scale requires significant investment in research and development, advanced equipment, and continuous optimization. Moreover, the integration of automated systems and real-time monitoring tools to handle large volumes of production often presents additional technical barriers. Economic challenges are also significant when it comes to scaling bioprocessing technologies. Although biotechnological approaches can lead to long-term cost savings by improving efficiency and reducing waste, the initial capital investment for large-scale infrastructure can be prohibitive. This includes the costs of setting up specialized production facilities, purchasing bioreactors, and ensuring the availability of necessary raw materials. Additionally, scaling up bioprocessing may require skilled labor and expertise in biotechnology, which can contribute to higher operational costs. There are also **regulatory hurdles** that must be overcome, as scaling up integrated food and bioprocessing often involves navigating complex and sometimes inconsistent regulatory frameworks across different regions. For businesses to successfully implement and scale these technologies, there must be strong regulatory alignment and clear guidelines to ensure food safety, environmental impact management, and consumer acceptance. Lastly, environmental concerns remain crucial in the scaling process. While bioprocessing offers promising solutions for sustainability, the large-scale implementation of these technologies must be carefully managed to avoid unintended environmental impacts, such as overuse of resources or unintended waste generation. Ensuring that the scaling process remains truly sustainable requires continuous innovation and adaptation to minimize the ecological footprint of bioprocessing systems. Balancing growth with environmental stewardship remains one of the key challenges for the successful implementation of integrated food and bioprocessing technologies.

10 The Vision for the Future

The vision for the future of integrated food and bioprocessing is centered around creating a more sustainable, efficient, and resilient global food system. As the world faces mounting challenges like population growth, climate change, and resource scarcity, the need for innovative food production methods has never been more critical. In the coming years, integrated food and bioprocessing technologies are expected to revolutionize the food industry by leveraging biotechnology to optimize production processes, enhance nutritional quality, and reduce waste. The integration of biological processes, such as fermentation and enzyme-based technologies, will become standard practice, helping industries produce food in a more resource-efficient manner while maintaining or improving food safety and quality. The future will see greater synergy between the food and biotech sectors, driving the development of new food sources, improved food preservation methods, and eco-friendly packaging solutions. Looking ahead, we can expect to see further advancements in plant-based and alternative protein products, which will contribute to reducing the environmental impact of animal agriculture. Additionally, innovations in automation, artificial intelligence (AI), and machine learning will streamline bioprocessing, improving efficiency and reducing costs across the food production chain. These technologies will also play a role in personalized nutrition, enabling the production of tailored foods that meet the specific health needs of individuals. The future of integrated food and bioprocessing will also be marked by a shift toward circular economies, where waste from food production is repurposed or upcycled into valuable by-products, such as biofuels, bioplastics, or animal feed. By focusing on sustainability, efficiency, and innovation, the food industry will move closer to meeting global food security demands while minimizing its environmental footprint, paving the way for a healthier and more sustainable future for all.

References

Niu, D., Zhang, M., Mujumdar, A. S., & Yu, D. (2024). Novel drying technologies for future animal-and plant-derived protein foods: Research progress, challenges, and potential applications. *Food and Bioprocess Technology*, 1–21.

Pang, C., Yin, X., Zhang, G., Liu, S., Zhou, J., Li, J., & Du, G. (2021). Current progress and prospects of enzyme technologies in future foods. *Systems Microbiology and Biomanufacturing, 1*, 24–32.

Index

A. S. Jatoi (ed.), *Integrated Food and Bioprocessing*,
https://doi.org/10.1007/978-3-031-87209-9

The manufacturer's authorised representative in the EU is Springer Nature Customer Service Centre GmbH, Europaplatz 3, 69115 Heidelberg, Germany. If you have any concerns regarding our products, please contact ProductSafety@springernature.com

Printed and bound by CPI Group (UK) Ltd, Croydon, CR0 4YY
30/09/2026
02225596-0001